COGNITIVE GENETICS

PETER SEIDLER

ARCTURUS

Cognitive Genetics is a Trademark of Cognigenics Inc.
Cover image: "Basal Ganglia" by Greg Dunn, 2025. Additional extrodinay brain images by Greg Dunn are throughout the book.

Eagle, Idaho 83616, publisher@arcturuspress.com
ISBN: 978-1-892654-42-7

COGNITIVE GENETICS

Preface

Welcome to *Cognitive Genetics*, a collection of interrelated writings examining the frontier of genetic engineering and RNA-based therapeutics for mental disorders. At the intersection of genetics and neuroengineering lies transformative potential—not just theoretical promise, but tangible treatments emerging now that point to an array of applications in medicine and a revolution in treatments including new forms of cognitive enhancement. This book elucidates pioneering advancements driving discoveries in Cognitive Genetics, offering insights into how these innovations are poised to address some of our most pressing health challenges.

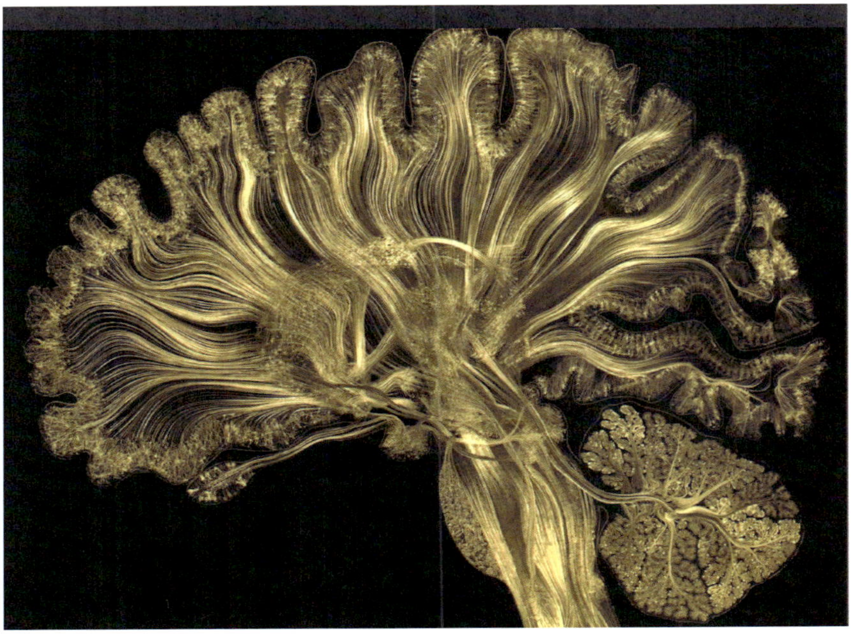

Self Reflected, Greg Dunn

Over the past decade, genetic engineering and artificial intelligence fields have accelerated, delivering revolutionary developments that promise to reshape our understanding and treatment of neuropsychiatric and neurocognitive conditions. Even calling this an "acceleration" understates the reality—we are already witnessing supportive scientific breakthroughs that were unimaginable just years ago. Integrating these technologies into clinical treatments holds the potential to move beyond symptom management, offering safe, genuine healing for neuropsychiatric and neurocognitive conditions.

To support the translation of this vital work into actual treatments, this text *Cognitive Genetics*, aims to bridge traditional psychotropic medications and the future of precision medicine. We highlight the role of gene therapy and include carefully curated lists and methodologies that provide glimpses into the language and determinants of this emerging field. I've found well-curated lists to be productive contemplative objects since they condense volumes of information into small, easily understandable segments and hierarchies. From these lists, we can swiftly comprehend crucial elements and determine priorities in this complex landscape.

Genetic Medicines for Neuropsychiatric and Neurodegenerative Disorders

As an executive at a pioneering neuroengineering gene therapy organization focusing on developing RNA therapies for neurological illnesses, I have firsthand experience with the scientific, creative process in the expanding field of Cognitive Genetics. A sizable and rapidly growing audience is interested in the development of RNA-based therapies, and rightfully so. The successes of research and development-driven enterprises in this space will help far more people than those in our immediate research and business communities.

Like many readers, there are individuals I deeply love who are right now experiencing severe mental disorders. I remember sitting with my college roommate during his first major depressive episode—watching helplessly as his brilliant mind turned against itself, trapped in cyclical thoughts that no amount of reason or care could penetrate. Twenty years later, he still struggles with the same fundamental issues, despite trying nearly every medication available. I write with vivid urgency, knowing more effective treatments are within reach. I believe some form of these therapeutics will very soon be able to address the growing suffering caused by these disorders as we explore and learn from scientists' knowledge and bridge the gap between cutting-edge research and the people it seeks to benefit.

Cognitive Genetics presents key factors surrounding gene therapy treatments, including RNA-based therapeutics such as short hairpin RNA and RNA interference, which allow for precise control of neuronal expression. These technologies and emerging tools enable targeted genetic interventions that may address underlying causes of a wide range of neuropsychiatric and neurodegenerative disorders. The resulting medicines and noninvasive delivery systems can bypass the blood-brain barrier, providing hope for a new era of precision therapies.

In *Cognitive Genetics,* we examine biomarkers to diagnose and monitor central nervous system illnesses. RNA biomarkers offer precise, non-invasive methods for early disease identification and treatment evaluation, opening doors for better patient outcomes. This is critical in an era when standard diagnostic methods significantly fall short of recognizing early-stage neurodegenerative disorders. We also explore the ethical and societal implications of these advances. As genetic therapies emerge, regulators must ensure that treatments are distributed fairly and used appropriately. This book reinforces the need to maintain high ethical standards while expanding our scientific understanding.

At the Threshold of a New Era in Mental Health

In creating *Cognitive Genetics*, I'm deeply grateful for the opportunity to collaborate over the last few years with incredible teams of researchers dedicated to advancing RNA neuromodulators. Their dedication and inventiveness have been a continuous source of inspiration. During a recent holiday, I watched my aunt struggle to recognize family members due to advancing Alzheimer's—the same brilliant woman who once taught me calculus now unable to follow a simple card game. Each such moment reinforces the urgency of this work and reminds me that behind every data point in a clinical trial is a human being and a family in pain.

This book aspires to bring together academic researchers, doctors, pharmaceutical teams, and others interested in the future of brain health and the development of cell therapies to modulate neural networks. It began as a series of articles I wrote this year, each tackling different aspects of Cognitive Genetics. As a result, you will notice recurring themes and ideas across chapters. Rather than editing out these repetitions, I've chosen to keep several to emphasize key facets and perspectives I believe are worth deepening. I hope these resonant elements strengthen the text's overall message and utility.

Cognitive Genetics describes scientific concepts, technological achievements, and potential applications with ideas likely to interest those far outside the scientific community. As we enter a new era of genetic treatments for mental health conditions, with each new discovery and increased "visibility" at the molecular level, we see the majesty of the display of interdependent dance of energetic complexity in coemergent brain–consciousness. Developing new therapeutics from this research, may help us in our drive to improve the greater human cognitive well-being and capacity.

In addition to my contributions, *Cognitive Genetics* contains fascinating articles from neuroscience pioneers Dean Radin and J. L. Mee. In the upcoming pages begins a generous foreword from Dean Radin. We gladly welcome you, to the future of cognitive well-being.

Peter Seidler | Princeton, NJ, 2024

Legal Disclaimer: This book is for informational and educational purposes only and does not provide medical, legal, or professional advice. The content is not a substitute for consulting licensed healthcare professionals, financial advisors, or legal experts. The authors and publishers are not responsible for any consequences resulting from the use of or reliance on the information in this book. Readers should seek proper professional advice before making any decisions based on its content.

Artificial Intelligence: AI tools support my writing process, saving time on repetitive tasks, quickly develop ideas I would not have had time to explore otherwise.

Foreword

With the rapid development of technologies that can precisely modulate genetics and epigenetics, *Cognitive Genetics* is emerging as an exciting new discipline focusing on the biological underpinnings of human cognition, behavior, and mental disorders.

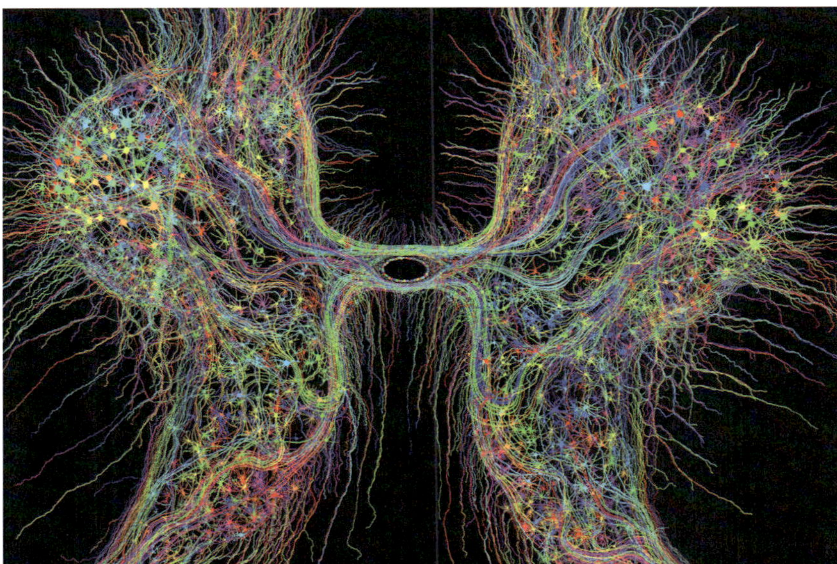

Grey Matter, Greg Dunn

Cognitive Genetics explores the intersection of genetics, neurobiology, psychological science, and psychiatric medicine. This book deepens our understanding of genetics' complex interplay with environmental factors and the potential applications of genetic engineering to fine-tune these elements for enhancing mental well-being. *Cognitive Genetics* is also the name of the scientific field seeking to understand how to use genetic engineering to modulate these factors to improve mental health. Topics of interest within the field include the full range of mental health disorders as well as the genetic underpinnings of extraordinary intelligence, memory, and prosocial behavior.

Cognitive Genetics' medical, ethical, and even evolutionary implications are profoundly important, on par with the civilization-shaping developments of industrialization, nuclear power, computers, and the Internet. Genetic engineering techniques like the 2020 Nobel Prize-winning development of CRISPR-cas9 are already used to treat previously intractable single-gene health disorders like Sickle Cell Disease, Hemophilia, Cystic Fibrosis, and Spinal Muscular Atrophy.

Peter Seidler's book surveys how neurogenetic engineering is being developed to treat mental health disorders. The leading edge of research in this emerging field is developing clinical genetic treatments for anxiety, depression, and mild cognitive impairment, as well as studying more complex challenges like schizophrenia, bipolar disorder, and neurodegenerative diseases like Alzheimer's.

The genetic architectures underlying cognitive functions like intelligence and creativity and the social and emotional traits that govern social behavior are polygenic, meaning they involve constellations of hundreds of interacting genes and epigenetic modulation. At the heart of this exploration lies the important question of how to leverage what we know about these factors to improve mental health and promote prosocial behavior and to do it in both scientifically sound and ethically responsible ways.

The Promise of the Cognitive Genetics Field

Many cognitive and emotional traits considered personally and socially positive, including intelligence, creativity, memory, compassion, and empathy, have significant heritable components. The same is true for traits that are considered antisocial. Our genetic inheritance does not determine these traits but shapes our potential, amplified or suppressed by our environment.

Beyond the first-line clinical applications for mental disorders, the advancement of Cognitive Genetics offers possibly revolutionary ways of encouraging prosocial behaviors—traits like empathy and cooperation. Understanding the genetic and epigenetic factors underlying these traits, developing ways to inspire more compassionate and cohesive societies, or, more immediately, methods for addressing persistent cultural or border disputes may be possible.

Of course, the development of such methods raises significant ethical issues that must be carefully considered. Using knowledge from *Cognitive Genetics* to go beyond treatment for neuropsychiatric and neurodegenerative disorders and to enhance "normal" cognitive functioning to foster prosocial behavior raises serious questions about human identity, autonomy, and the very nature of our species. The ability to develop "enhanced humans" is also enormously seductive. Such efforts are already underway, at least in laboratory models, to create human babies that are engineered not only to avoid known genetic diseases but also to select other desirable traits.

In addition, while the promotion of prosocial behavior would appear to be benign provided that ethical factors are carefully considered, it may also be possible to promote antisocial behavior, for example, in combat soldiers. Historically, the feared Norse warriors known as the "berserkers" were renowned for their ferocity and seemingly supernatural strength in battle. Some have speculated that their frenzied states were produced through a combination of psychoactive drugs and ceremonial practices. But could similar states also be brought about through genetic interventions?

Such questions are not just abstract concerns. They are possibilities when considering how rapidly *Cognitive Genetics* develops and how that knowledge may be applied outside clinical medicine.

A Balanced Approach

Because *Cognitive Genetics* is likely to bring a range of promises and perils, the best approach is to maintain humility. As with any significant advancement in fundamental science, Spider-Man was advised to remember:

With Great Power Comes Great Responsibility

Maintaining great responsibility means collaborating across scientific disciplines and with psychologists, sociologists, ethicists, and policymakers. Such dialogues may help society navigate safely through this new minefield.

As you engage with the material in Peter Seidler's book, I encourage you to consider the implications of *Cognitive Genetics* and how we can harness its potential to enhance human flourishing and health while avoiding the seductions of creating "superhumans."

To learn more, please see references in *Selected Sources, by Chapter.*

Dean Radin, Ph.D. | Boise, ID, 2024

Acknowledgments

My heartfelt thanks go to the remarkably skilled team whose collaboration made this book possible.

I owe special gratitude to Tracy Brandmeyer, Troy Rohn, and Fabio Macciardi, who not only embody how to approach the intricacies of Cognitive Genetics, but also offered invaluable support as we explored these concepts. Their scientific insight and patient guidance transformed challenging material into accessible knowledge.

I'm deeply thankful to J. L. Mee, whose thoughtful conversations generated concepts for this book, and whose approach to scientific exploration continues to be an inspiration.

A particular thank you to Janet Fletcher, a gifted book editor, for excellent structural advice that significantly improved the flow of this work. I would like to extend my thanks to Maia Robbins for her line editing, which enhanced the readability.

I also want to extend loving thanks to John and Susan Johnson, for their encouragement and wisdom.

Finally, my greatest gratitude and thanks to Chananya and our children Elena, Theo, and Orion who have shown remarkable patience during the late nights and distracted weekends this project demanded. I bow to you and will be making pancakes in the morning.

Contents

Preface v

Foreword By Dean Radin ix

Acknowledgments xiii

Introduction By J. L. Mee xxiii

Part One: Social

1 **A Quest for Innovative Treaments** 3
A focus on creating and developing novel RNA-based therapeutics for neurocognitive disorders and cognitive optimization.

2 **A New Class Of Genetic Therapeutics For Mental Health** 7
Introduces cutting-edge RNA neuromodulators for mental health disorders. By Dean Radin

3 **Precision & Promise: RNA Neuromodulation** 11
A spotlight on RNA therapeutics as potentially transformational treatments for neuropsychiatric diseases.

4 **Safety, Efficacy, And Patient Outcomes** 17
Compares the efficacy and safety of RNA therapeutics to established mental health treatments.

5 **Mechanics To Genetics: Remembering Technological Epochs** 23
Impact of technological advancements from the Industrial Revolution to the Genetic Age, triumphs, challenges, and ethical considerations.

6 **Quantum AI Safety and Cognitive Capital** 27
What happens when AI starts running on quantum computers?
By J. L. Mee

7 **Broadening Perspectives** 31
The role of shRNA in addressing Substance Use and Opioid Use Disorders.

Part Two: Individual

8 **The Future of RNA Therapies For Enhancing Cognitive Health** 37
Considers how RNA therapy may help improve mental function and avoid cognitive deterioration.

9 **The Science of Stress Reduction** 44
RNA therapies have the potential to reduce the neurological effects of chronic stress. Investigates the underlying causes of stress.

10 **Memory Decline and Cognitive Impairment** 50
Highlights potentially critical role of RNA therapies combating memory loss and cognitive decline, offering treatment possibilities.

11 **Neurocognitive Stress Cascade: Neuromodulator Treatment** 55
NSC is a neurochemical system mobilizing the body against threats, involves brain regions, hormones, and feedback loops. Research reveals RNA-based neuromodulators' potential.

12 **Next-Generation Personal Change** 65
Combining cognitive neuro-engineering technology with human potential development techniques can expand cognitive capacity. By J. L. Mee

13 **Cognitive Resilience Continuum** 69
Cognitive functioning is mapped along a spectrum of durable and measurable states of cognition and the baseline cognitive function for each state.

Part Three: Brain

14 **Long-Lasting Genetic Therapeutics For Debilitating Neurological Conditions** 75
Intranasal gene-editing treatments can target specific neural networks. By Dean Radin

15 **Cognitive Longevity & Hippocampal Neuron Hyperactivity** 79
RNA treatments may lower neuronal hyperactivity, particularly in the limbic system circuits improving cognitive lifespan.

16 **Artificial Intelligence and Genetic Medicine** 87
AI is revolutionizing RNA therapies by enhancing genetic research and improving precision of RNA treatments.

17 **An Inquiry Into Brain Coherence** 93
The synchronization of electrical activity across different brain regions is crucial for cognitive and emotional functioning.

18 **Innovative Longevity Biomarkers For CNS Disorders** 103
RNA biomarkers can enable much earlier, more accurate diagnoses of CNS illnesses by detecting disease-specific RNA changes.

19 **Overcoming the Blood-Brain Barrier** 107
Explores innovative methods for effectively delivering RNA therapeutics across the blood-brain barrier to treat neurological disorders.

20 **Comparing Methods of Gene Editing in the Design and Production of Neuromodulators** 111
Innovative methods for delivering RNA therapeutics across the blood-brain barrier to treat neurological disorders.

Part Four: Neuron

21 **Precise Genetic Targeting in Neuropsychiatric Disorders** 119
Discusses genetic editing techniques that are in development, such as CRISPR and shRNA, for precise treatment of mental disorders.

22 **How RNA Therapies Target Neurons for Modulation** 123
RNA therapies deliver molecules binding to specific mRNA sequences.

23 **Harnessing RNA Therapeutics to Address Neuroinflammation** 127
Examines potential of RNA therapies in treating neuroinflammation, a critical factor in neurological diseases.

24 **Synthetic Biology** 131
The transformative potential of synthetic biology as a groundbreaking general-purpose technology, predicting a profound impact on society. By J. L. Mee

25 **Wide-Ranging Benefits of 5-HT2a Receptor Reduction** 135
Reducing 5-HT2a receptor density is a key RNA therapy target for mental health, linked to stability, cognition, and neuroprotection, offering hope for resistant conditions.

Part Five: Pillars of Innovation

26 **Scientific Review** 145
Contextualizing the "Pillars of Innovation" in RNA Therapeutics for Neuropsychiatry and Neurodegenerative Conditions.

27 **Paradigm Shift** 147
Article: "Intranasal Delivery of shRNA to Knockdown the 5-HT2a Receptor Enhances Memory and Alleviates Anxiety."

28 **Blueprint for Tomorrow's Medicine** 149
Article: "Treatment with shRNA to knockdown 5-HT2a in primary cortical neurons. receptor improves memory in vivo and decreases excitability."

29 **The Dawn of Precision Medicine for Mental Health** 151
Article: "Genetic Modulation of the HTR2A Gene Reduces Anxiety-Related Behavior in Mice."

30 **Executive Summary** 153

Appendix

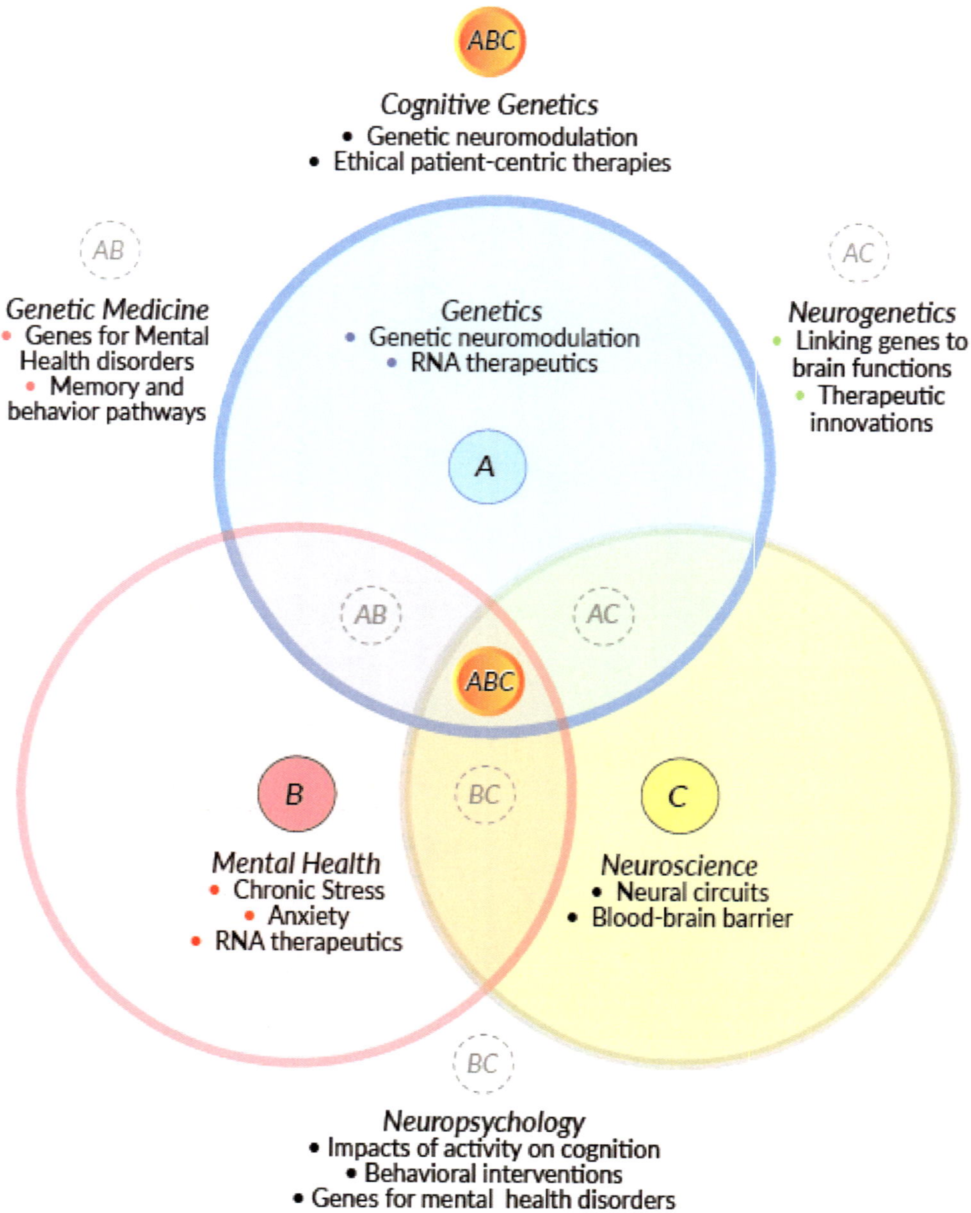
ABC
Cognitive Genetics
• Genetic neuromodulation
• Ethical patient-centric therapies
AB
Genetic Medicine
• Genes for Mental Health disorders
• Memory and behavior pathways
Genetics
• Genetic neuromodulation
• RNA therapeutics
AC
Neurogenetics
• Linking genes to brain functions
• Therapeutic innovations
A
AB
AC
ABC
B
BC
C
Mental Health
• Chronic Stress
• Anxiety
• RNA therapeutics
Neuroscience
• Neural circuits
• Blood-brain barrier
BC
Neuropsychology
• Impacts of activity on cognition
• Behavioral interventions
• Genes for mental health disorders

Interdisciplinary Foundation for Cognitive Genetics

As illustrated on the opposite page, the convergence of three fundamental disciplines—Genetics, Mental Health, and Neuroscience—form the foundation of Cognitive Genetics and drive development of innovative therapeutics.

Genetics

- Targeted gene modulation technologies
- Epigenetic regulation mechanisms
- Precision delivery systems for genetic material

Mental Health

- Chronic stress and anxiety disorders
- Cognitive decline and memory impairment
- Treatment-resistant depression

Neuroscience

- Neural circuit function and modulation
- Blood-brain barrier crossing technologies
- Neuroplasticity mechanisms

Where these fields overlap, important subfields emerge:

Genetic Medicine (Genetics + Mental Health)

- Biomarkers for early detection
- Personalized treatment approaches
- Genetic risk factor identification

Neurogenetics (Genetics + Neuroscience)

- Gene expression profiles in neural tissues
- Heritable factors in neural development
- Receptor-specific targeting mechanisms

Neuropsychology (Mental Health + Neuroscience)

- Brain-behavior relationships
- Cognitive assessment methodologies
- Neural correlates of psychological states

At the intersection of these disciplines is the emerging field Cognitive Genetics, merging precision neuroengineering with patient-focused therapies. The foundation for next-generation treatments, targeting specific neuronal receptors in brain regions driving the neuropsychiatric and neurodegenerative conditions.

Introduction

Over the past fifty years, rapid advancements in science and technology have given rise to groundbreaking fields such as genetic engineering and artificial intelligence (AI). These innovations promise transformative benefits but also present substantial challenges.

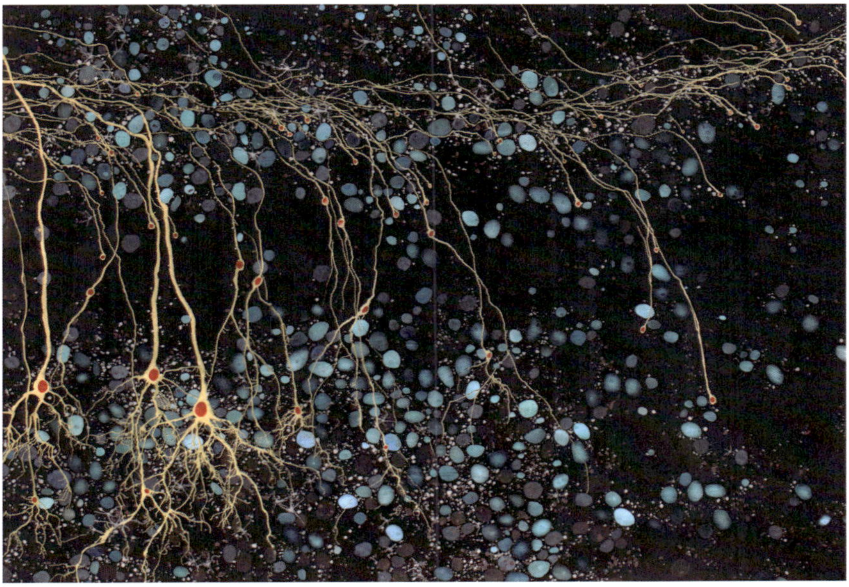

Neurogenesis, Greg Dunn.

The advent of the Internet has democratized access to information, yet it has simultaneously raised concerns about social isolation. Similarly, the growth of AI forces humanity to confront and potentially transcend its cognitive limitations. This progression is far more significant than simply another step in the evolution of robotics.

The Need for a New Approach

Predicting the future of our world in fifty years is daunting due to the accelerated pace of scientific breakthroughs. The Genetic Age holds promise, but also considerable risk without a collective improvement in humanity's collective awareness, perception, cognitive abilities, and ethical understanding. Our wisdom needs to catch up with our technical proficiency.

Traditional methods and practices for achieving higher states of consciousness are slow and unable to keep pace with technological advancements. Thus, there is a pressing need for radical, disruptive thinkers to foster a more conscious future for humanity. Consciousness engineering is increasingly seen as a vital component of scientific evolution.

Renowned historian Dr. Yuval Noah Harari emphasizes the transformative impact that small genetic and neuronal changes have historically had on human evolution. Today, humanity has the technical capability to influence states of consciousness. Genes linked to cognitive capacity have been identified, and microbiological tools for modifying them have been successfully demonstrated in laboratories.

CRISPR was first successfully used in 2023 to enhance cognition in normal adult mice, boosting memory by 104%. CRISPR was programmed to modify neuron electrodynamics to improve the animals' attention, enabling them to form stronger memories. Attention in mice, proxy for awareness in humans..

These discoveries could potentially enhance human cognitive abilities, leading to significant advancements in arts, sciences, and overall human progress, though the full extent of their impact is yet to be understood.

A New Frontier in Cognitive Genetics

The potential to elevate human cognitive capacity represents a pivotal moment in human civilization's progress.

Emerging *Cognitive Genetics* technologies can be harnessed to

The Gene Editing Revolution

"The discovery of the century." – Bloomberg

"DNA editing will remake the world." – Wired

"The gene-editing tool's potential to upend science is dizzying." – Vox

"CRISPR will change the world forever." – ScienceAlert

"The potential of gene editing is enormous." – TIME

"This revolutionary gene-editing tool could change the world." – NBC News foster the best outcomes for human endeavors. Advances in genetic engineering and cognitive science are poised to revolutionize mental health treatments, offering solutions that go beyond symptom masking to provide true healing without side effects.

Revolutionizing Consciousness and Cognitive Evolution

The power to influence human evolution through subtle genetic and neural modifications has reached new heights. With the identification of genes linked to cognitive abilities and proven strategies to modify them, we now have the tools to create entirely new states of consciousness. These groundbreaking developments promise to elevate human cognitive potential, paving the way for unprecedented advancements in knowledge, creativity, and the sciences.

Cognitive Engineering

Investing in human cognitive enhancement alongside AI development is imperative to avert potential risks associated with the imbalance between human and artificial intelligence. By expanding human cognitive potential, we can ensure that humanity remains at the forefront of technological advancement, capable of understanding, guiding, and controlling robust AI systems.

Genetic Engineering

Genetic engineering can enhance cognitive faculties such as mental acuity, awareness, and intelligence. Revolutionary laboratory techniques like RNA editing and CRISPR have decoded the previously unfathomable mechanics of gene interactions. This development is a pivotal achievement in the history of biological research, with the potential to transform human civilization akin to the steam engine and that of the transistor.

Genetic Neuropsychology

A new branch of science is emerging that improves human cognition by genetically optimizing brain functioning. Genetic neuropsychology combines neuroscience, cognitive psychology, and genetic engineering to treat neurological disorders. These disorders usually involve imbalances in various brain regions, and by normalizing the unbalanced areas with CRISPR, a disorder's behavioral and cognitive effects can be mitigated. This new science is made possible by rapidly evolving new forms of genetic editing, which can dynamically deliver genome edits to most brain regions. This highly flexible technology can modify any neuron, gene, or RNA transcript, representing a foundational scientific breakthrough.

Treatable neurological conditions include attention deficit disorder, obsessive-compulsive disorder, post-traumatic stress disorder, anxiety, depression, psychosomatic issues, memory concerns, Alzheimer's disease symptoms, and mild cognitive impairment.

Genetic neuropsychology can also be used for cognitive enhancement to raise conscious awareness, decrease inattention, reduce mind-wandering, lower craving, sharpen mental focus, increase concentration, enhance mindfulness, increase intuition, and foster emotional balance.

Neurological Treatment Market

According to World Bank projections, over the next twenty years, $6 trillion in global health expenditures will shift from treating symptoms with medicine to addressing causes with DNA editing. The World Health Organization estimates cognitive impairment and other neurological disorders account for 10% of global health issues, which would represent $600 billion in costs.

Cognitive Capital

We are witnessing the emergence of a new kind of wealth—cognitive capital. This new form creates value in many applications, but perhaps its greatest potential for value creation lies in cognitive capital. Cognitive capital is the source of all cognitive talents, including creativity, focus, compassion, motivation, and mental understanding, and it is the measure of an individual's free, unbound conscious awareness in the present.

The integration of scientific breakthroughs with consciousness evolution is what is required. Genetic engineering and cognitive science advancements offer unprecedented opportunities to enhance human cognitive capacities, address mental disorders, and foster a higher state of consciousness. Innovative researchers stand at the forefront of this transformative movement, leveraging genetic technologies to revolutionize mental health treatment and elevate human potential. In the following chapters of this book, Peter Seidler writes about the manybemerging aspects of gene therapies for neuropsychiatric and neurocognitive health.

J. L. Mee | Stuart, FL, 2024

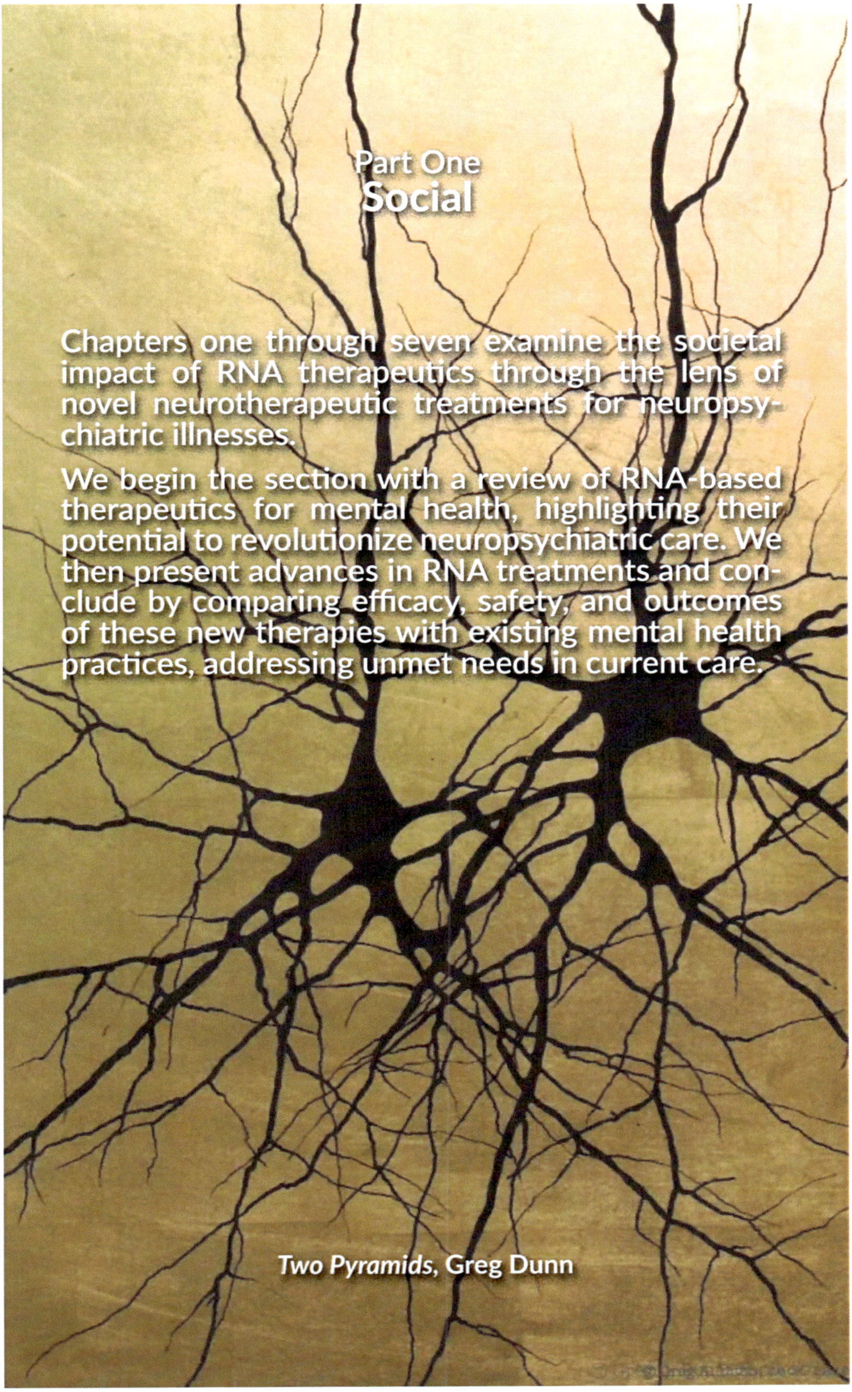

Part One
Social

Chapters one through seven examine the societal impact of RNA therapeutics through the lens of novel neurotherapeutic treatments for neuropsychiatric illnesses.

We begin the section with a review of RNA-based therapeutics for mental health, highlighting their potential to revolutionize neuropsychiatric care. We then present advances in RNA treatments and conclude by comparing efficacy, safety, and outcomes of these new therapies with existing mental health practices, addressing unmet needs in current care.

Two Pyramids, Greg Dunn

1. A Quest for Innovative Treatments

As researchers dedicated to this field, many of us witness first-hand the devastating impact of neuropsychiatric conditions on our loved ones. Many of us know our family member's growing distress as early Alzheimer's symptoms manifest alongside intense anxiety, a combination conventional medications barely touch.

Genetic medicine. Rendering.

Watching someone you love repeatedly ask the same questions, each time with increasing fear in their eyes, creates an urgency that transcends academic interest. It's these deeply personal experiences that drive our commitment to developing better solutions with the precision and efficacy that current treatments lack.

Practical Applications and Effectiveness in Clinical Settings

The beauty of shRNA molecules lies in their precise targeting—they bind to specific genes' messenger RNA (mRNA), effectively inhibiting the expression of particular genes and reducing the synthesis of proteins associated with illness. This therapy operates with precision at the molecular level. In laboratory animal studies, this approach yields precise modulation of neural networks affected in conditions such as anxiety and memory impairment.

Precise gene targeting is the cornerstone of these therapies. When treating neuropsychiatric diseases, imprecise targeting may result in unwanted side effects. Well-designed shRNA can specifically target certain proteins, such as neuronal receptors, while minimizing effects on other brain pathways.

Encouragingly, preclinical research in animal models demonstrates substantial reductions in anxiety and significant enhancements in memory. These improvements occur without any discernible adverse behavioral effects—a promising indicator for potential human applications.

Evidence from Preclinical Research

Preclinical research provides the foundation for predicting clinical impact in human patients. Researchers have conducted thorough in vitro studies using human neurons alongside in vivo experiments with rodent models. The human neuron culture experiments consistently demonstrate that shRNA treatments effectively reduce the protein expression of targeted genes. When targeting the 5-HT2a neuronal receptor, therapy led to significant reductions in the protein receptor and a corresponding decrease in electrical excitability—changes that correlated with decreased anxiety behaviors and improved memory.

Electrophysiological studies further validate the molecular mechanism of shRNA, strengthening preclinical findings. For instance, multi-electrode array (MEA) recordings in cultured neurons have shown that treatment with shRNA targeting the 5-HT2a receptor leads to changes in neural firing patterns that correspond to reduced neuronal hyperactivity. These findings align with a growing body of research linking decreased hippocampal hyperactivity to enhanced memory retention and cognitive function in animal models.

Patient Compliance and Noninvasive Delivery Methods

Researchers are developing innovative RNA therapies using the nose-to-brain intranasal delivery route, which elegantly bypasses the blood-brain barrier. This intranasal approach represents a noninvasive and patient-friendly alternative to conventional methods. For central nervous system disorders, this delivery method may offer distinct advantages over traditional oral or injectable medications. Through the intranasal route, therapeutic compounds travel directly to the brain via the olfactory pathways and trigeminal nerves.

The Rostral Migratory Stream (RMS), prominent in rodents, plays a more subtle but important role in human adults than previously thought. Together, these three anatomical pathways effectively deliver therapeutic cargo to neuronal populations within the central nervous system. Preclinical studies have validated this delivery method's efficacy. With proper molecular design, RNA molecules reach targeted brain regions, producing therapeutic benefits. The simplicity and noninvasive nature of this approach are likely to enhance patient compliance—a crucial element in treating chronic conditions. Additionally, the long-lasting effects of RNA therapies reduce administration frequency, further improving compliance and the overall patient experience.

Absence of Unintended Effects: Safety Profile

A primary concern with gene therapies like shRNA is the potential for off-target effects leading to unintended outcomes. Fortunately, These RNA therapies have demonstrated a favorable safety profile in preclinical studies. The specificity of shRNA targeting significantly reduces the probability of off-target gene silencing by ensuring only the intended gene and selective protein are affected.

Early findings are particularly encouraging—even when administered at doses exceeding those anticipated for human trials, test animals showed no adverse behavioral effects or physiological changes such as weight loss. This safety buffer is compelling evidence that these therapies will likely be well-tolerated by humans. The absence of adverse reactions, combined with precise targeting of specific neural circuits, suggests that RNA therapies may offer a safer alternative to many current pharmacological treatments.

Potential Applications for a Broad Range of Conditions

While scientists are currently focused on developing RNA therapeutics for anxiety and mild cognitive impairment, the underlying technology holds promise for treating a diverse spectrum of neuropsychiatric and neurodegenerative illnesses.

RNA Neuromodulators: Expanding Framework & Applications

The emerging framework for RNA neuromodulators represents a comprehensive approach to understanding, developing, and implementing RNA-based treatments for neurological conditions. As research advances and our understanding deepens, scientists can build diverse pipelines of RNA therapeutics—including siRNA, mRNA, and antisense oligonucleotides—addressing a wide range of diseases with increasing precision and efficiency.

The versatility of RNA neuromodulators enables precise selection and adjustment of numerous genetic targets, extending their utility to conditions such as Alzheimer's disease, depression, schizophrenia, Huntington's disease, and ADHD. The ability to personalize these neuromodulatory interventions for specific psychiatric and cognitive conditions opens additional avenues where therapies might be customized to enhance cognitive capacity even in healthy individuals.

Positive Effect and Effectiveness

These RNA treatments show remarkable promise for treating neurocognitive and neuropsychiatric illnesses. Their development represents progress in RNA therapeutics for mental health, with transformative potential stemming from promising preclinical outcomes, innovative delivery mechanisms, and strong safety profiles.

A key factor in this success is the evolving framework supporting development of RNA neuromodulators that enables efficient design, optimization, and delivery of treatments. This research framework's integrated components—from RNA sequence design to understanding neurological pathways and preclinical testing—allow rapid iteration and refinement, ensuring each neuromodulator achieves the best design.

As these medicines progress through clinical trials, the benefits of this RNA neuromodulator framework become increasingly evident. For scientists, the ability to target multiple conditions or receptor pathways with these neuromodulators improves safety while reducing development time. This research-informed approach decreases costs and accelerates the journey from discovery to clinical trials, offering hope for improved treatment options.

Every time I visited my colleague's mother in the memory care facility—once a brilliant physician herself—and observe her struggle to recognize her own son, I'm reminded that our timeline for innovation isn't just a scientific roadmap but a race against deterioration for millions of individuals. For these patients and their healthcare practitioners, the expanding potential of RNA neuromodulators brings not just academic optimism but tangible hope—the kind families cling to when conventional treatments have failed them.

2. A New Class of Genetic Therapeutics for Mental Health

By Dean Radin

The personal and societal burdens of debilitating mental health disorders, including the growing prevalence of mild cognitive impairment with associated anxiety, motivate the development of novel therapeutics.

NG2 Flare, Greg Dunn. Phenomenon involving NG2 glia, precursor cells.

Background

There are currently no FDA-approved pharmaceutical treatments for mild cognitive impairment (MCI), nor for MCI with associated anxiety. Importantly, these two disorders often go hand in hand (Orgeta et al., 2022). As the elderly population increases among developed nations, MCI and anxiety already affect 10% to 15% of those 65 and older, and because MCI is often a transitional stage into the more serious diagnosis of dementia (Anderson 2019), there is a demand for new and more effective therapies.

Many drugs that are currently used to treat chronic anxiety and depression were approved in the late 1980s. This includes selective serotonin reuptake inhibitors (SSRIs, like Zoloft) and similar approaches based on norepinephrine (SNRIs, like Cymbalta). These drugs have proven to be useful treatments for some patients, but besides continuing concerns about a wide range of serious side effects (largely attributed to the systemic delivery of the drug, which affects 15 subtypes of serotonin receptors), they are not intended to improve memory, and in some cases may exacerbate memory problems (Anagha et al., 2021). Meanwhile, drugs like Donepezil (brand name Aricept) which are used to treat dementia, are not intended as anxiolytics, and may likewise result in paradoxical adverse reactions (Hakimi et al., 2020).

In search of more effective treatments to manage chronic anxiety, depression, and cognitive decline, including MCI, there has been a resurgence of interest in the use of serotonergic hallucinogenic compounds (i.e., psychedelics, like psilocybin), which act as serotonin 2A receptor (5-HT2aR) agonists (Garcia-Romeu et al., 2022). Clinical studies with these compounds continue to show promising efficacy, but their hallucinogenic properties, which may play an important therapeutic role (Raj et al., 2023; van den Berg et al., 2022), will probably limit their widespread acceptance. This has led to the development of psychedelic analogs, such as molecules similar to LSD that are said to not produce hallucinations (Lewis et al., 2023). Whether such approaches will prove to be therapeutically efficacious and safe remains to be seen.

A Novel Approach

Given the demand to develop an effective treatment for MCI and anxiety, what is desirable is a therapeutic that (1) precisely targets only the 5-HT2a receptors in the brain, (2) can be delivered easily and noninvasively directly to the brain, (3) shows superior efficacy in treating MCI and anxiety, (4) lasts many months with each treatment, (5) has no undesirable physiological or behavioral side effects, (6) has no discernable toxicological or safety concerns, and (7) would be cost comparable to existing pharmaceutical treatments.

A solution is described in an article that successfully addresses each of the first six elements in mice and rats, with a similar electrophysiological profile demonstrated in human iPSC neurons in vitro. (Rohn et al., 2023).

After discussions with pharmaceutical consultants and genetic reagent manufacturers, we have reason to believe that the seventh desirable element, a relatively low cost, is also achievable due to the large potential patient need. *Cognigenics* first tested their intranasal genetic editing approach using CRISPR-cas9, in mice, to permanently modify the 5-HT2a receptor, and found it to be highly effective in improving memory and reducing anxiety However, it it is unlikely, at least at this early phase of the development of neurogenetic therapeutics, that a DNA-based treatment would be approved for a fairly common mental health disorder. This motivated our current work, where we have demonstrated that we can achieve equally effective results, as a temporary treatment, with an RNA edit.

Our approach, which relies on the RNA-induced silencing complex (RISC) pathway, allows for an extremely precise form of RNA interference editing. To study the level of precision and help address concerns about the possibility of off-target edits, we searched for potential matches of our editing sequence throughout the entire mouse genome, and found only one—the intended target that results in silencing production of the 5-HT2a receptor. Our method is thus a precise 5-HT2a antagonist (and not a broad stroke agonist, like psychedelic compounds), which in standard rodent behavioral tests leads to both a significant reduction in anxiety and a highly significant improvement in memory.

Takeaway Messages

At this preclinical stage of our R&D, based on our experimental studies, we have high confidence that our approach is an effective treatment for MCI and anxiety in mice and rats, with no evidence for off-target, safety, or toxicology concerns. Whether this method of treating mental health will be equally effective in humans or not, require clinical trials, which we are actively working toward achieving. If those trials are successful, then RNA editing may help to create a new class of highly efficacious and precision therapeutics for a broad range of psychiatric disorders, beginning with MCI and anxiety.

3. Precision & Promise: RNA Neuromodulation

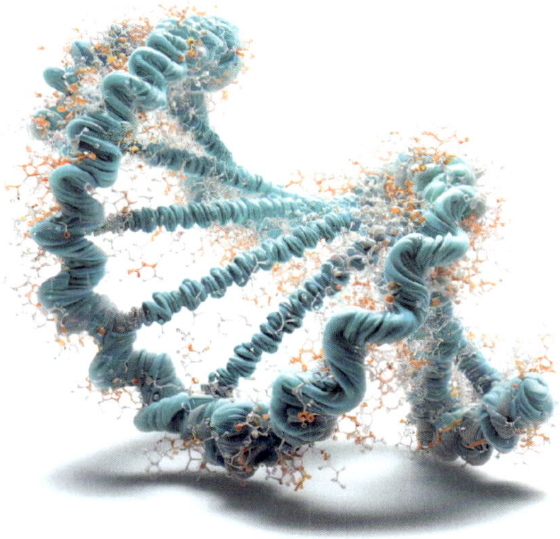

shRNA molecule for treating neuropsychiatric conditions. Rendering.

The precision of shRNA treatment opens remarkable new avenues for treating neurocognitive and neuropsychiatric conditions. Having witnessed a close friend's decade-long struggle with treatment-resistant depression—cycling through medication after medication, each with its own constellation of side effects—I've developed a profound appreciation for the need for more targeted therapies. This method's tailored approach enables precise regulation of neuronal protein targets implicated in various illnesses. A precision therapy focuses on correcting the dysfunctional state while leaving off-target proteins in their normal form undisturbed. This level of specificity is critical for medicines targeting complex brain disorders, where precision impacts treatment outcomes and patient quality of life.

Reduction in Disease Biomarkers

The promise of shRNA treatment extends well beyond symptom management to address the underlying molecular causes of neurocognitive and neuropsychiatric disorders. If these preclinical findings translate successfully to humans and accurately address the foundation of these illnesses, shRNA treatment has the potential to fundamentally alter disease trajectories. This approach may not only alleviate symptoms but also modify disease progression at the molecular level, offering hope for conditions previously considered intractable.

The disease-modifying capacity of shRNA treatment highlights one facet of its revolutionary nature. Beyond preventing disease progression by calming neuroinflammation, the therapy's ability to alter gene expression raises the possibility of restoring cognitive and mental function to a natural state of equilibrium. This characteristic of shRNA treatment may transform therapeutic goals, shifting from palliative care to a more curative strategy.

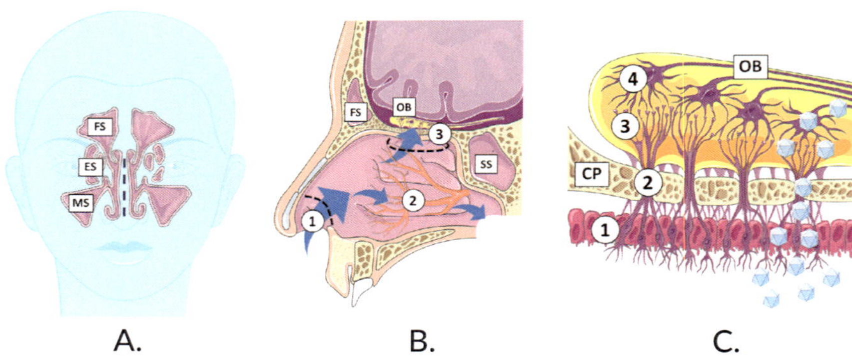

Figure: Schematic anatomy of the nose and the olfactory system] (Figure contains adapted clipart from Servier Medical ART [https://smart.servier.com/]) A. Anterior projection of the paranasal sinuses. The dashed line represents the location of the nasal septum. B, View of the lateral wall of the main nasal cavity. The three areas of the nose are indicated by numbers and separated by dashed lines: (1) nasal vestibule, (2) respiratory region, and (3) olfactory region. The arrows depict the direction and potential targets of AAVs: the central nervous system via the olfactory bulb, the respiratory system via the nasopharynx, and systemic spread via blood vessels in the respiratory region. C, Depiction of the olfactory epithelium and bulb as well as passage of AAVs into the central nervous system. (1) Exposed dendrites of olfactory neurons reach into the nasal cavity. (2) Axons of olfactory neurons penetrate the cribriform plate and (3) form spherical clusters, called glomeruli, which connect to (4) the mitral cells within the olfactory bulb. AAVs can directly be transported via this way into the central nervous system. Abbreviations. FS, frontal sinus; ES, ethmoid sinus; MS, maxillary sinus; SS, sphenoid sinus; OB, olfactory bulb; CP, cribriform plate. **(Gadenstaetter, AJ et al., 2022)**

Long-Term Benefits

These RNA neuromodulators potentially offer significantly longer duration of action than traditional pharmaceuticals. The sustained nature of shRNA therapy will transform how we approach chronic neurocognitive and neuropsychiatric conditions. By providing sustained action, shRNA therapy elegantly addresses the challenge of patient compliance. This persistent effectiveness can dramatically improve the therapeutic experience and supports a more holistic approach to patient care that prioritizes quality of life.

I still remember visiting my former colleague, once a brilliant teacher, now struggling with early-onset Alzheimer's. Despite her disciplined approach to medication, she frequently forgot doses or accidentally doubled them—a cruel irony of the very disease she had spent her career studying. The prospect of longer-acting treatments that reduce this burden brings not just scientific excitement but profound human relief.

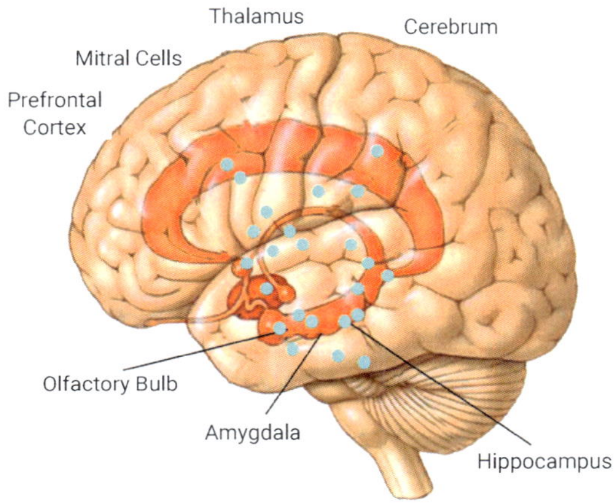

Drug reaching various parts of the brain through olfactory/trigeminal pathways, cerebrospinal fluid, and the Rostral Migratory Stream. American Journal of Nuclear Medicine and Molecular Imaging (2020)

Adeno-associated viruses (AAVs) represent the most commonly employed vectors for targeted gene delivery, and extensive study has led to the approval of multiple gene therapies. The intranasal route of vector application offers several advantages over traditional administration methods. In addition to targeting local tissue like the olfactory epithelium, it provides minimally invasive access to various organ systems, including the central nervous system and the respiratory tract (Gadenstaetter, AJ et al., 2022).

Minimized Side Effects

The exquisite selectivity of shRNA treatment provides a substantial benefit in reducing side effects. Side effects from off-target engagement are a prevalent issue with many traditional CNS medications. For example, there are fifteen different serotonin receptor subtypes, and SSRIs can potentially impact every single one, both systemically and within the CNS. However, a designed shRNA that specifically targets only one of these receptors, such as the 5-HTR2a, offers a clear advantage.

The selectivity of these neuromodulators decreases the risk of systemic effects and interactions with medications for comorbid conditions, as well as unwanted CNS side effects. This is especially valuable for elderly patients, who frequently experience polypharmacy—the simultaneous use of multiple medications, often to manage complex health conditions, typically involving five or more drugs concurrently. Improved patient tolerance and adherence, resulting from reduced side effects, can lead to better treatment outcomes and a more favorable risk-benefit ratio.

Addressing Potential Hurdles

Various obstacles impede the advancement of shRNA-based therapies. The primary goal is to facilitate reliable and precise delivery of therapeutic RNA, bypassing the blood-brain barrier (BBB) while ensuring selectivity within the complex brain environment.

The BBB poses a significant challenge as it selectively inhibits the passage of most substances from the blood to the brain, impeding targeted application of therapeutic agents. Researchers are examining advances in nanotechnology and molecular engineering to overcome this obstacle. Scientists are developing liposomes, nanoparticles, and other cutting-edge delivery techniques to increase the effectiveness of transporting shRNA across the BBB. These carriers can be designed to target specific brain regions while protecting shRNA from degradation, ensuring that therapeutic molecules remain intact and functional once they reach their neuronal targets. Achieving consistent distribution and stability is critical for the success of shRNA treatments.

Mitigating Off-Target Effects in RNA Therapy

Short hairpin RNA therapy's effectiveness hinges on its high specificity. This is achieved through precise design of the RNA sequence to match the target mRNA and careful delivery to specific neuronal sites. The targeted approach of RNA neuromodulators offers significant advantages.The RNA cargo is designed to regulate only the intended gene in specific cell types, minimizing interference with other genes and reducing the risk of unintended effects elsewhere in the genome. Compared to traditional medications, which interact with already-produced proteins, shRNA therapy intervenes at the mRNA level, preventing protein production altogether. This approach allows for more precise control of gene expression.

Despite its high selectivity potential, RNA therapeutics face challenges in the form of off-target effects. These can occur when RNA recognizes unintended mRNA targets, potentially leading to unwanted effects. To address this, researchers employ various solutions to improve accuracy.

Bioinformatics tools and genome editing techniques are utilized to evolve shRNA designs. These advanced methods help in predicting and evaluating potential off-target interactions, thereby enhancing the precision and safety of shRNA therapy. Scientists use peer-reviewed applications to predict potential off-target actions for any given designed shRNA, aiding in the evaluation of interactions between shRNA and the genome.

Ensuring the safety and efficacy of shRNA therapy involves rigorous in vitro and in vivo analysis. This thorough evaluation of interactions between shRNA and the genome is crucial for demonstrating the therapy's specificity and safety profile.

Clinical Translation

Regulatory considerations play a significant role in the development of shRNA therapies. Obtaining regulatory clearance depends on demonstrating specificity, proving minimal off-target effects, and showing an overall favorable safety profile. These stringent requirements ensure that only well-designed and thoroughly tested shRNA therapies reach clinical applications, prioritizing patient safety and treatment efficacy.

To successfully transition from the laboratory to the clinic, researchers must address scientific and technological challenges, as well as ethical, regulatory, and societal concerns. Prior to testing shRNA therapies in human subjects, these treatments must undergo rigorous preclinical evaluation to determine their safety and efficacy. Early engagement with regulatory bodies during the development process is crucial to ensure that gene therapies such as RNA interference meet the stringent safety and efficacy standards required for clinical trials and eventual market approval.

Transparent communication with the public and patient advocacy organizations will enhance trust and promote better understanding of RNA-based therapies. It is crucial to educate all stakeholders—including scientists, physicians, regulatory agencies, and patient advocacy organizations—about the benefits of a collaborative approach. This collaboration is necessary to safely and effectively integrate new therapies like shRNA into clinical practice. These partnerships will ensure that the development and application of shRNA treatments are governed by ethical principles, patient needs, and scientific rigor, leading to improved outcomes.

The application of shRNA therapy represents an innovative approach to addressing neuropsychiatric disorders, offering potential solutions for conditions previously considered beyond the reach of conventional medicine. As I look at photographs of my aging parents on my desk while writing this chapter, I'm reminded that we're not just tackling technological challenges—we're racing against time for millions who need these treatments now. Behind every scientific advance is the profound human hope of transforming lives through the careful integration of gene-based medicines into clinical practice.

4. Safety, Efficacy, and Patient Outcomes

Over one billion people globally experience neurological disorders, including anxiety, depression, and cognitive decline. Behind this staggering statistic are countless personal stories—like my own uncle who, despite decades of conventional treatment for his bipolar disorder, still experiences debilitating cycles that leave his family constantly bracing for the next crisis.

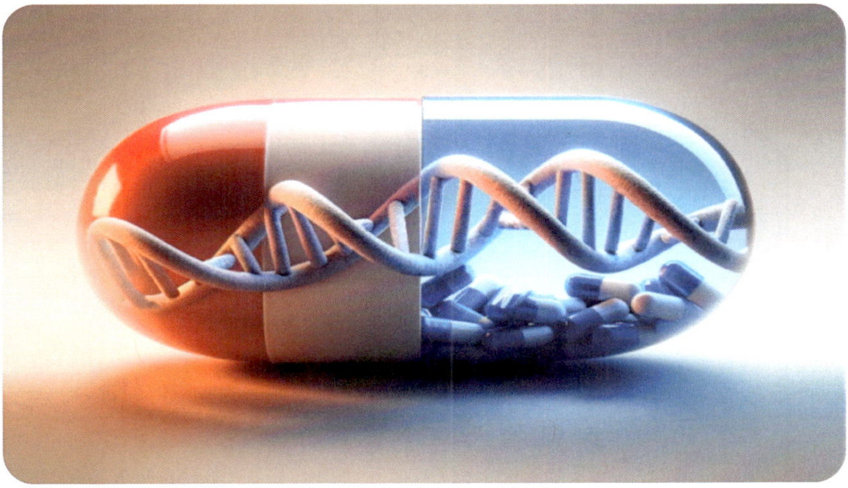

Transformation of pharmaceuticals to genetic medicines. Rendering.

Healthcare systems face substantial hurdles in addressing this reality. Despite their significant limitations in efficacy and tolerability, traditional psychotropic medicines, such as SSRIs and benzodiazepines, have been widely used historically for therapy. RNA neuromodulators have emerged in recent years as a viable

"Our focus is on leveraging the most advanced genetic tools to offer new hope for those who have not found relief with traditional therapies."

– Dr. Tracy Brandmeyer, Neuroscientist

alternative. These treatments provide focused interventions that have the potential to dramatically enhance patient outcomes. This chapter compares RNA-based approaches and traditional drugs, specifically focusing on their effectiveness, safety, and impact on patient results.

Mental disorders have a substantial impact on worldwide health systems. Traditional psychotropic drugs, such as selective serotonin reuptake inhibitors (SSRIs) and benzodiazepines, have been the standard of care for these problems.

Traditional Psychotropic Medications

For decades, standard psychotropic medicines have been the primary treatment choice for mental health issues. Each of these medications has a distinct mode of action. SSRIs, such as Prozac, Zoloft, and Lexapro, work by increasing serotonin levels in the brain, which can alleviate symptoms of depression and anxiety. Benzodiazepines, such as Xanax and Valium, increase the activity of the gamma-aminobutyric acid (GABA) neurotransmitter system. This can have an anxiolytic action and help to reduce tension.

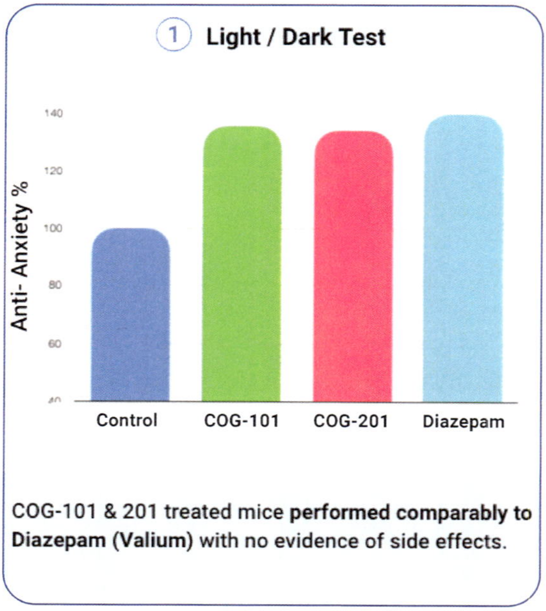

Preclinical Results

I still remember watching a dear friend meticulously organize her weekly pill organizer—a rainbow of medications that often left her feeling, as she describes it, "like a zombie." This common experience highlights the fundamental limitation of traditional approaches, paticularly given the rapid growth in these conditions.

Comparative Effectiveness

Researchers have shown promising preclinical results for RNA neuromodulators targeting neuropsychiatric and neurodegenerative disorders. Studies in animal models have demonstrated long-lasting and highly significant reductions in anxiety and improvements in memory performance.

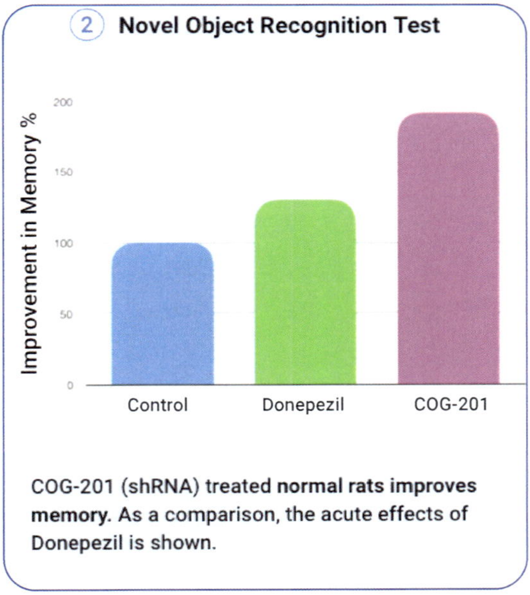

Preclinical Results

In contrast, traditional psychotropic drugs have varying levels of efficacy, for some, the respite they bring is fleeting. According to research findings, many individuals who initially respond to SSRIs eventually stop therapy due to adverse effects or diminishing benefits over time.

Given the current situation, there is an urgent need for new therapeutic solutions that are more effective and dependable. RNA therapies show considerable potential to address this requirement. The precision targeting offered by RNA neuromodulators allows for intervention at the genetic level, potentially correcting the underlying mechanisms of disease rather than merely masking symptoms.

Patient Outcomes and Quality of Life

The most important and essential factors in assessing available treatments for mental health issues ultimately come down to patient outcomes and quality of life. To improve patient adherence and general satisfaction, RNA therapies may provide particular

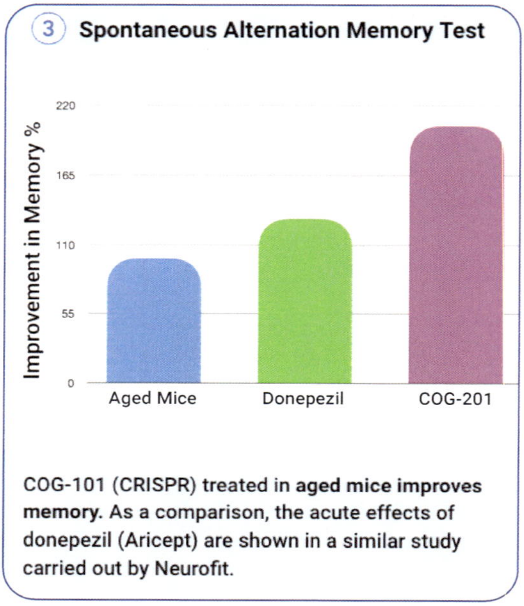
3
Spontaneous Alternation Memory Test
Improvement in Memory %
220
165
110
55
0
Aged Mice
Donepezil
COG-201
COG-101 (CRISPR) treated in aged mice improves memory. As a comparison, the acute effects of donepezil (Aricept) are shown in a similar study carried out by Neurofit.

Preclinical Results

advantages. One main benefit is that RNA-based medicines may be administered in a single dose, or a dose every three to four months. This simplifies treatment plans and is likely to improve patient compliance.

Walking through a memory care facility recently, I observed an elderly woman struggling as a nurse attempted to administer her daily medications. "I don't have any problems," she insisted repeatedly, unaware of her advanced dementia. The nurse explained to me later that this scene repeats daily, creating distress for both patient and caregiver. The prospect of treatments requiring administration only a few times yearly would transform this daily struggle into a rare occurrence.

Further, the specificity of RNA neuromodulators ensures that therapeutic effects are delivered to targeted brain circuits, reducing the possibility of off-target effects and increasing treatment effectiveness. Patients may experience fewer side effects and continuous symptom relief from this precision targeting, improving their quality of life and treatment outcomes. By contrast, standard psychotropic drugs frequently require ongoing administration, which often results in problems with adherence and treatment resistance. As a result, we often see reduced quality of life because of enduring adverse effects.

Moreover, it has been estimated that roughly 40% of patients with clinical depression or anxiety do not respond to traditional treatments including SSRIs. These patients are often referred to as "treatment resistant," and this population would benefit greatly from nonconventional therapeutics including RNA interference compounds. For these individuals, RNA neuromodulators represent not just an alternative treatment but potentially their first experience of meaningful relief.

Future Directions and Innovations

Researchers are actively exploring several promising future directions and innovations in RNA-based therapeutics. As RNA therapy delivery techniques evolve, improvements in precision and efficacy are becoming evident. These advancements allow for better targeting of specific neuronal receptors, increasing the success rates of treatments.

The rapid progress in RNA-based treatments is driven by a confluence of factors. First, the unmet need for effective therapies in mental health, especially for conditions resistant to traditional pharmaceuticals, motivates substantial investment in innovative approaches. Additionally, shRNA-based therapies provide significant opportunities by addressing the root causes of neural dysregulation without the side effects of current treatments. These

> "Research shows targeted gene editing in the hippocampus can significantly improve memory and reduce anxiety in animal studies."
>
> – Fabio Macciardi, MD, Ph.D.

therapies can deliver both temporary and permanent modifications to neural circuits, creating personalized treatment pathways.

With rising global demand for new treatments, particularly for disorders like Alzheimer's and anxiety-related cognitive decline, the field presents a fertile landscape for innovation. The search for better treatment options isn't merely academic for me—it's deeply personal. Every time I visit my mother-in-law, who struggles with medication-resistant anxiety that has dramatically narrowed her world, I'm reminded of how urgently we need these advances. Watching someone you love lose their independence not to cognitive decline itself but to the paralyzing fear that accompanies it creates an immediacy that transcends simple scientific curiosity.

Researchers are leveraging these advancements to explore broader applications in neurogenetic medicine, offering hope for those suffering from conditions currently untreatable by existing methods. As these RNA neuromodulators move closer to clinical application, they carry with them the potential to fundamentally transform our approach to treating disorders of the mind—shifting from management to modification, from coping to correction, and ultimately, from merely extending life to truly enhancing its quality.

5. Mechanics to Genetics: Remembering Technological Epochs

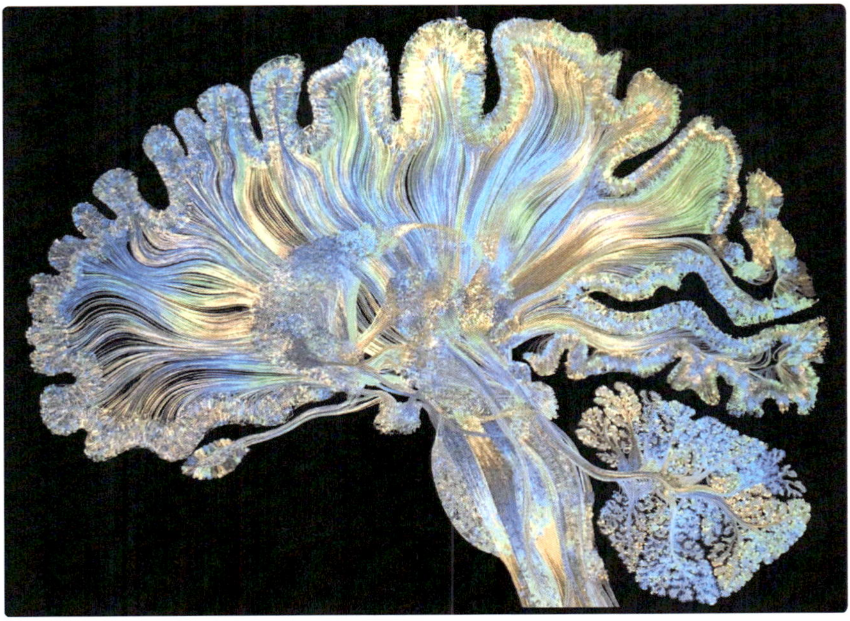

Self Reflected, Greg Dunn

A sequence of breakthrough technical revolutions has fueled humanity's unrelenting progression of technological developments. Every period, from the beginning of the Industrial Revolution to the power of the Genetic Age today, has brought forth cataclysmic upheavals that have transformed economies, civilizations, and the fabric of human experience and consciousness. This chapter reviews the significance of these revolutionary advances, triumphs, challenges, and the ethical implications of the potential to produce life itself.

Laying the Foundations of Modernity

The Industrial Revolution, which began with steam power, signaled the shift from agricultural to industrial civilization. James Watt created the steam engine in the late nineteenth century, providing the framework for mechanized transportation and industrial progress. As factories replaced traditional modes of production, laborers relocated to cities in pursuit of jobs, hastening urbanization.

During this period, industrial capitalists and the emergence of the working class created new social problems and exacerbated economic inequality.

Expansion, Electrification, and Global Reach

The Industrial Age witnessed the electrification of towns, industries, and transportation networks, expanding on the gains made during the Industrial Revolution. The advent of the electric generator and motor enabled the generation of electrical power, which led to large-scale industrialization and urbanization.

The rise of consumer culture and advertising was spurred by mass manufacturing techniques, such as Henry Ford's assembly line invention, which made consumer products more affordable. However, social disparities worsened throughout this time, even though the Industrial Age considerably raised living standards and economic prosperity. The rising gap between the wealthiest industrialists and the working class highlighted the need for social reforms and worker rights in business success. Furthermore, the environmental consequences of industrialization were increasingly apparent, bringing new challenges in terms of air and water pollution.

On the other hand, the electrification of cities and the widespread use of electric appliances and lighting altered family life. These enhancements reduced home obligations and introduced new ways of life. While the Manufacturing Age demonstrated how inventive people can be and how they can utilize technology to transform the world, it also showed the need to deal with the unintended consequences of development.

The Digital Revolution Reshapes Society

The Information Age began in the latter part of the twentieth century, signaling a shift in how we engage, communicate, and interact with the world around us. The microchip, also known as the integrated circuit, became the modern era's enabling technology, significantly increasing processing power and allowing the advent of personal computers and the internet. The digital revolution has shifted the economic focus away from manufacturing to knowledge-based industries and services.

The arrival of the internet and mobile communication devices enabled instantaneous global connectivity and information access, connecting individuals in previously unheard-of ways. This opened up new opportunities for trade, education, and international contact, but it also highlighted concerns about cybersecurity, privacy, and the digital divide between those with and without access.

The Information Age demonstrated digital technology's transformative potential while highlighting ethical and management responsibilities. While we welcomed the digital era's benefits and opportunities, we faced challenges such as misleading information, cyberbullying, and the consequences of automation on jobs.

Engineering Life and Redefining Possibilities

With the advent of the Genetic Age, we are ready for yet another significant transformation, with R&D-oriented RNA corporations leading the development of genetic medicine. The capacity to decode and manipulate the code of life has ushered in a new age known as the Genetic Age.

Gene-editing methods such as CRISPR have created new opportunities in agriculture, health, and other areas. Teams that embrace these breakthroughs set the standard for creating RNA therapeutics and genetic medicine and are at the forefront of this transition.

RNA R&D companies demonstrate the potential of the Genetic Age by applying targeted genetic treatments to cure mental health disorders and neurological illnesses. Using RNA-based medicines and genetic neuroengineering, these firms aim to address the underlying genetic causes of ailments such as anxiety, memory loss, depression, and Alzheimer's disease with unprecedented accuracy and efficacy.

The Genetic Age reflects a paradigm change from treating symptoms to addressing the underlying causes of illnesses at the molecular level. However, the ability to design life itself has tremendous ethical consequences. Concerns about bioethics, equal access to new technologies, and any implications for the natural environment must be addressed.

The situation requires us to seize the potential of this revolutionary moment while adhering to the highest ethical standards and practicing responsible stewardship. Revolution necessitates a deep awareness of its societal implications and an unflinching dedication to overcoming the challenges it offers.

New Perspectives and Considerations

Each technological revolution has had a significant environmental impact, from the Industrial Age's air pollution and greenhouse gas emissions to e-waste and carbon-intensive energy needs. As we approach the Genetic Age, we must emphasize sustainable behaviors while mitigating the possible environmental implications of genetic engineering and biotechnology.

Ethical Governance and Regulation

Technological innovations have regularly exacerbated economic and social gaps while promoting social justice and fair access, marginalizing groups. The benefits of genetic medicine and biotechnology must remain inexpensive and available to all people, regardless of financial status or geographic location, in the Genetic Age.

The potential to influence the most fundamental aspects of existence needs the development of solid ethical frameworks and regulatory oversight. Governments, scientists, and bioethicists must collaborate to develop standards that support responsible innovation while preserving human rights, preventing abuse, and upholding the sanctity of life.

Fostering Public Dialogue and Education

The potential use of genetic engineering to increase human capabilities raises concerns about human identity and the need for safeguards to protect dignity.

Due to the complexities of the Genetic Age, open discourse and public education are essential to ensure that informed decisions are made. Scientific researchers, legislators, and educators can engage the general public in discussions about the potential risks, benefits, and ethical dilemmas linked with genetic technology. This promotes trust and openness.

Navigating the Future with Wisdom and Foresight

From the steam-powered machinery of the Industrial Revolution to the genetic miracles of the current day, enormous changes are continuously reshaping human civilization. These developments reflect the spirit of human ingenuity and our capacity to imagine.

6. Quantum AI Safety and Cognitive Capital

By J. L. Mee

What Happens When AI Starts Running on Quantum Computers?

Microprocessor power has been the driving force behind technological advances for decades. The accelerated progress in physical sciences has created innovations that have reshaped industries, while the human sciences have yet to keep pace with an equivalent force.

Quantum Computer

Before the advent of genetic techniques like CRISPR, there was no counterpart to Moore's Law in the humanities, where advancements and learning could grow exponentially. As a result, a widening "wisdom gap" has emerged, with human cognitive and social development lagging behind the innovations in physical sciences. This growing disparity raises profound challenges for future governance and the responsible development of powerful technologies like artificial intelligence (AI).

AI advancements are accelerating at an unprecedented rate, and the potential risks associated with its misuse grow exponentially. The challenges related to maintaining control over AI systems are outpacing humanity's current capacity to anticipate and manage these technologies safely. With policymakers and business leaders often driven by short-term goals, the long-term consequences of unleashing unregulated, highly sophisticated AI could have dire repercussions for society. A key concern is the delegation of increasingly powerful tools to individuals who may not fully understand or appreciate the ethical implications, making the need for caution and foresight ever more pressing.

Phase Change

The arrival of quantum computing in AI could be a tipping point. The transition of AI from conventional von Neumann architectures to quantum systems could exponentially increase its processing power. In contrast to classical computers, which process information in binary (0s and 1s), quantum computers leverage quantum bits (qubits), allowing for the superposition of multiple states simultaneously. This increased computational capacity means quantum AI could solve problems previously considered intractable. However, this surge in AI power may exacerbate the imbalance between machine intelligence and human cognitive abilities, pushing the limits of human oversight and control.

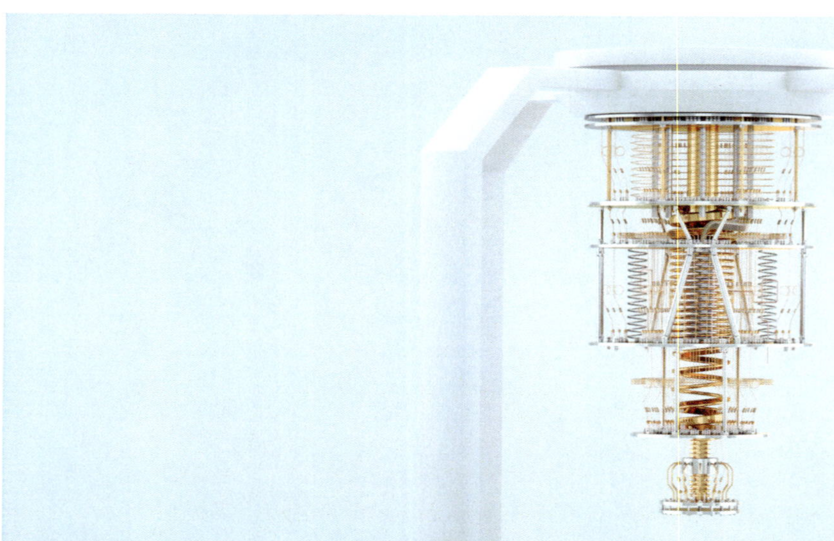

Quantum Computer

If society continues prioritizing developing more intelligent AI systems without proportionate investment in human cognitive enhancement, we may face a future where AI surpasses human decision-making abilities. Such an imbalance could lead to unprecedented risks, with AI making critical decisions autonomously and potentially without the necessary human context or ethical considerations. The fear is that the rapid advancements in AI, fueled by quantum computing, could lead to AI systems capable of manipulating human systems, behaviors, and even entire economies without the ability for humans to intervene effectively.

Cognitive Capital

There must be a parallel investment in enhancing human cognitive capacities to address this growing risk. The idea of cognitive capital—our ability to reason, innovate, and make decisions—is now more critical than ever. Developing technologies that can expand human cognitive skills will ensure that we remain capable of guiding the trajectory of AI development and harnessing its power safely and ethically. Doing so can maintain a balance where humans stay in control, guiding AI systems toward beneficial outcomes rather than unintended consequences.

Psychedelic research has opened up intriguing possibilities for enhancing human cognitive abilities. Studies have shown that compounds like psilocybin induce expanded consciousness, promoting creativity, mental flexibility, and even deeper self-awareness. These effects suggest that psychedelics could play a role in human cognitive enhancement by unlocking latent mental potential. Such findings have sparked interest in further exploring the use of psychedelics as tools for cognitive enhancement, a field that is now intersecting with advancements in genetic neuroengineering.

Enhancing Cognitive Function

Genetic neuroengineering, particularly through techniques like short hairpin RNA (shRNA) and CRISPR, presents a promising frontier for enhancing human cognition. These advanced technologies allow for precise modulation of gene expression, enabling targeted adjustments in neural circuits that govern cognitive functions.

By focusing on genes related to memory, attention, and decision-making, it is possible to enhance cognitive processes in ways that may help individuals keep pace with the rapid advancements in artificial intelligence. Such genetic modifications could lead to sustained improvements in mental acuity, creativity, and problem-solving, offering a counterbalance to the increasing sophistication of AI systems.

Investing in research and development for human cognitive enhancement is not only a strategic necessity but may also be an existential imperative. By enhancing human cognitive capital, we can ensure that we retain the ability to understand, regulate, and direct the future trajectory of AI technologies. This dual-track approach, focusing on both AI development and human enhancement, is critical for safeguarding humanity's future in an era of rapid technological change.

Biotech startups, for example, are at the forefront of pioneering genetic neuroengineering techniques to enhance human cognition. Their intranasal shRNA therapies targets specific brain regions to regulate neural activity, showing promise for treating conditions like anxiety and memory loss. This technology could be adapted to enhance cognitive function in healthy individuals, offering a new frontier in human cognitive enhancement. By focusing on mental health treatment and cognitive enhancement, companies like *Cognigenics* are paving the way for a future where humans can keep pace with AI-driven advancements.

In summary, the convergence of quantum computing and AI presents an unprecedented societal challenge. If we continue to develop AI systems without enhancing human intelligence, we risk creating an imbalance that could have far-reaching consequences. However, by investing in cognitive enhancement technologies—whether through psychedelics, genetic neuroengineering, or other leading-edge approaches—we can ensure that humanity remains in control of its technological destiny. A balanced approach to technological progress, one that enhances both machine intelligence and human cognition, will allow us to harness the full potential of AI while mitigating the risks associated with its misuse.

7. Broadening Perspectives

Is there a Role for shRNA in Addressing Substance Use, and Opioid Use Disorders?

Substance Use Disorder (SUD) and Opioid Use Disorder (OUD) represent some of the most pressing public health challenges today. These conditions, affecting millions globally, are deeply entwined with dysregulated neural circuits, leading to cycles of dependence, relapse, and societal burden.

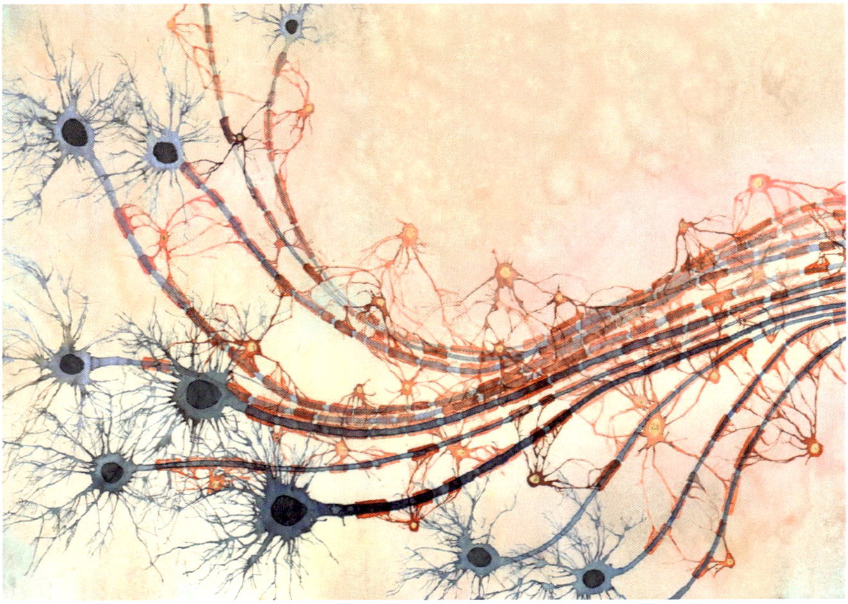

Myelination, Greg Dunn

Historically, treatments have focused on symptom management or substitution therapies, but they fall short of addressing the underlying neural dysregulation. What if we could go deeper—directly addressing the molecular and genetic drivers of these disorders? Enter RNA neuromodulators, a transformative genetic therapy offering unparalleled precision in treating SUD and OUD through their neurological mechanics. What if we could go deeper—directly addressing the molecular and genetic drivers of these disorders? Enter shRNA, a transformative genetic therapy offering unparalleled precision in treating SUD and OUD through their neurological mechanics.

Understanding the Neurobiology of SUD and OUD

SUD and OUD arise from the hijacking of the brain's reward pathways. Neural circuits linking the prefrontal cortex, hippocampus, and amygdala become hyperactivated, reinforcing compulsive drug-seeking behaviors. The hippocampus is particularly pivotal, anchoring environmental triggers and cravings in long-term memory. Disruptions in neurotransmitter systems—especially dopamine, serotonin, and glutamate—exacerbate this dysfunction, creating a feedback loop of craving and use.

Current pharmacological treatments like methadone or buprenorphine blanket these pathways but lack specificity, often causing side effects or dependency themselves. A precision-focused approach is needed. This is where short hairpin RNA (shRNA) may offer a revolutionary solution.

shRNA: Precision Engineering for Neural Circuits

shRNA acts as a molecular scalpel, silencing specific gene expressions associated with SUD and OUD. This technology targets the 5-HT2a receptor, a key player in serotonin modulation and neural excitability. The dysregulated activity of this receptor fuels cravings and reinforces compulsive behaviors in SUD and OUD. By silencing 5-HT2a activity, shRNA may break this cycle, reducing craving intensity and the emotional impact of addiction-related cues.

Many clinicians describe addiction as "memories that the body can't forget," capturing precisely what makes current treatments so limited—they attempt to manage a condition deeply encoded within neural circuits without addressing those circuits directly. The ability of shRNA to precisely target and modulate overactive neural circuits offers hope for long-lasting recovery.

Transformative Intranasal Delivery

Accessibility and patient compliance are crucial in addiction treatment. Intranasal delivery platforms for RNA neuromodulators bypass invasive procedures like brain injections or intravenous therapy. By leveraging the olfactory pathways, this method ensures that therapeutic shRNA reaches key brain regions such as the hippocampus and amygdala without crossing the blood-brain barrier systemically.

This non-invasive, patient-friendly delivery method promises significant advantages. It eliminates the stigma and discomfort associated with traditional interventions and offers a single-dose, long-lasting solution that fosters adherence and empowers recovery.

Addressing Unmet Needs in SUD and OUD Treatment

Despite substantial investment, the relapse rates in SUD and OUD remain alarmingly high. Existing treatments often fail due to their systemic side effects, dependency risks, or inability to address the root causes of addiction. The shRNA approach may offer sustained efficacy through long-term silencing of addiction-driving genes, safety and specificity by minimizing off-target effects, and no dependency, as it avoids substituting one addiction for another.

Building a Case for shRNA in Addiction Medicine

Expanding research to include SUD and OUD treatments represents both a strategic and ethical imperative. There is significant overlap between the neural mechanisms of addiction and the areas where early research indicates genetic neuromodulation medicine directly applies. Strategic collaborations with addiction research centers and clinical partners are necessary to accelerate the pathway to clinical trials. Public health organizations have expressed growing concern about addiction treatment efficacy, suggesting enthusiasm for innovative approaches that address genetic underpinnings.

Pioneering Ethical and Accessible Solutions

Companies creating these therapies must prioritize affordability, safety, and ethical distribution from preclinical studies to market deployment. This commitment extends to addressing the stigma surrounding SUD and OUD, fostering a recovery paradigm rooted in dignity and empowerment.

New Chapters in Addiction Treatment

RNA neuromodulator technology heralds a new era in addiction medicine. By addressing the genetic roots of SUD and OUD, this approach transcends traditional symptom management, offering genuine healing and hope. For the millions affected by these disorders, shRNA may unlock the possibility of long-term recovery and restored potential.

In support groups across the country, families affected by addiction share stories of lost potential, strained relationships, and perpetual worry. The promise of RNA neuromodulators isn't merely scientific advancement; it's the possibility of restoration for countless individuals and families trapped in cycles that current medicine cannot break.

This is not merely innovation—it is an imperative to rewrite the narrative of addiction and recovery. Through science, compassion, and precision, researchers are poised to transform lives and redefine the boundaries of possibility in addiction treatment.

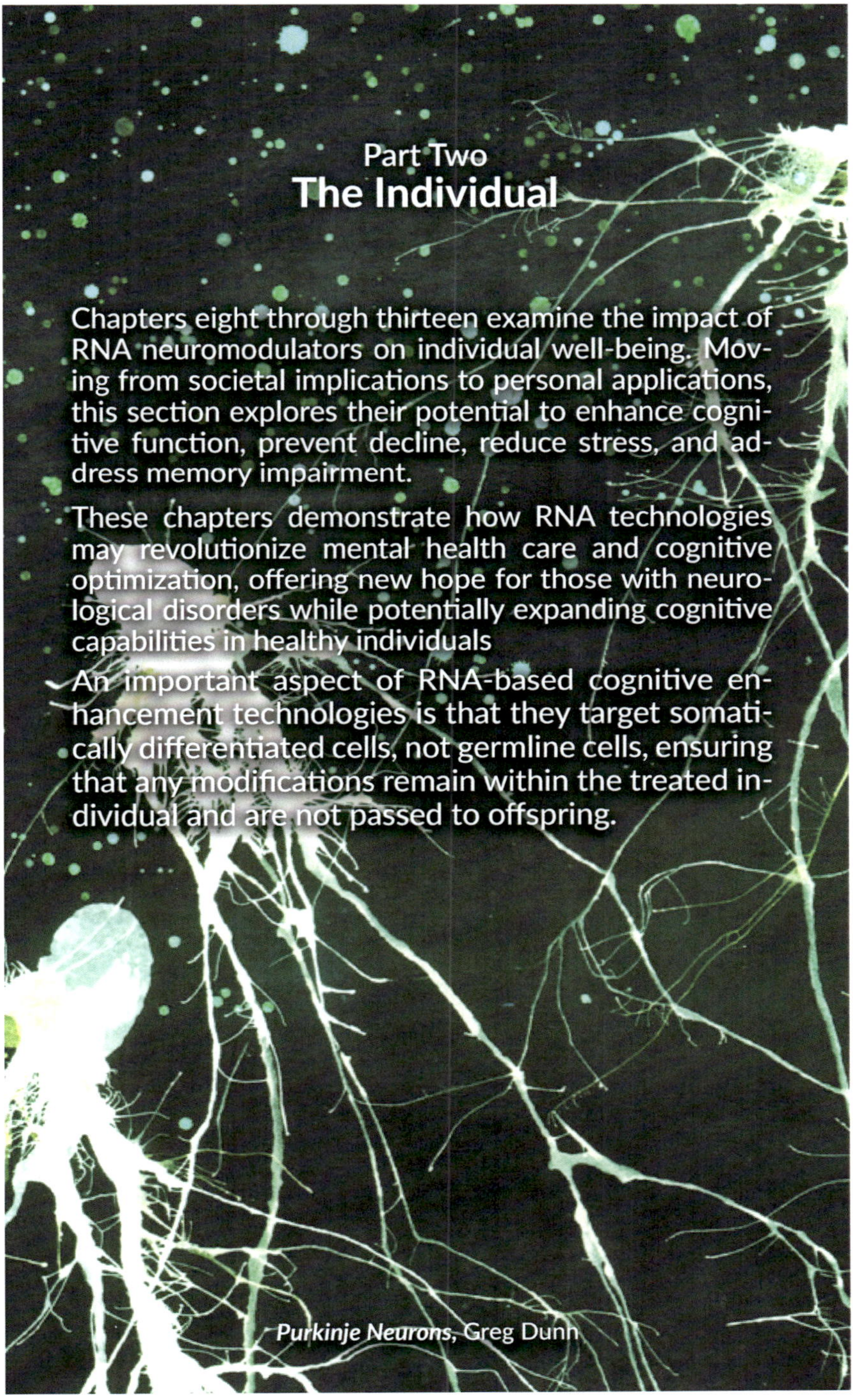

Part Two
The Individual

Chapters eight through thirteen examine the impact of RNA neuromodulators on individual well-being. Moving from societal implications to personal applications, this section explores their potential to enhance cognitive function, prevent decline, reduce stress, and address memory impairment.

These chapters demonstrate how RNA technologies may revolutionize mental health care and cognitive optimization, offering new hope for those with neurological disorders while potentially expanding cognitive capabilities in healthy individuals

An important aspect of RNA-based cognitive enhancement technologies is that they target somatically differentiated cells, not germline cells, ensuring that any modifications remain within the treated individual and are not passed to offspring.

***Purkinje Neurons*, Greg Dunn**

8. The Future of RNA Therapies for Enhancing Cognitive Health

A steady stream of biotechnology breakthroughs using gene therapy offers new approaches to improving mental health and cognitive abilities. While the medical community anticipates the clinical availability of RNA therapies for mental health, promising research breakthroughs are generating optimism. This growing interest extends to both those with neurodegenerative disorders and those in good health who wish to enhance their cognitive abilities.

5-HT2a neuronal receptor. Rendering.

RNA neuromodulators employ the body's gene expression control systems to selectively target genes involved in various physiological and pathological conditions. The selectivity of these approaches offers a significant advantage over traditional pharmaceuticals, which can have a wide variety of unanticipated side effects. Potential applications include addressing chronic neuropsychiatric conditions and boosting cognitive abilities in those without pre-existing health problems.

Current risk assessment frameworks for RNA-based therapies involve rigorous pre-clinical toxicity screening, dose-escalation studies, and long-term monitoring protocols. Researchers are particularly focused on evaluating off-target effects, immune responses, and potential impacts on related neural circuits. This multi-layered safety approach reflects the precision required when modifying gene expression in neural tissue.

This chapter investigates RNA treatments, including potential benefits, ethical issues, and aspirations for improved mental health.

Enhancing Synapse Formation and Cognitive Abilities with RNA Therapy

As we move forward, the applications of RNA-based therapies extend beyond treating disorders to optimizing brain function in healthy individuals—enhancing memory, learning, and overall cognitive performance. The ability to precisely target specific brain circuits with minimal side effects represents a groundbreaking advancement in mental health and neuroengineering.

Further research will focus on understanding the long-term effects of RNA therapies, including safety, dosage optimization, and the potential for personalization based on individual genetic profiles. The ultimate goal is to refine these treatments so they can be tailored to improve specific cognitive abilities or address particular neurological conditions without disrupting other brain functions.

Looking at realistic timeframes, RNA-based therapies for severe neurological conditions may enter early-stage clinical trials within the next one-to-two years, while applications for cognitive enhancement in healthy individuals likely remain 7-10 years from initial human studies. This progression reflects the regulatory pathway, which typically prioritizes applications for medical conditions over enhancement purposes. The technology's transition from bench to bedside will likely occur in stages, with treatments for well-defined neurological disorders paving the way for broader applications.

The implications of enhancing human intelligence through RNA therapies are vast. By boosting synaptic plasticity and promoting neural regeneration, we could unlock new possibilities for learning and memory, even in later stages of life. As these therapies progress through clinical trials, they may revolutionize our approach to mental health, cognitive enhancement, and neurodegenerative diseases, offering solutions that are both effective and sustainable.

RNA neuromodulators present a promising future in enhancing cognitive capacities and treating mental health disorders. With ongoing research, we are likely to see significant advancements in how these treatments can improve brain function, leading to a new era in neuroscience and genetic medicine.

Neuroplasticity, RNA Therapy for Cognitive Enhancement

Neuroplasticity—the brain's ability to adapt and reorganize in response to learning and experiences—forms the foundation for cognition and memory retention. RNA therapy has been shown to enhance neuroplasticity, potentially increasing the brain's ability to undergo synaptic changes, known as synaptogenesis.

RNA neuromodulators can accelerate synaptogenesis, strengthen existing neural connections, and promote the formation of new synapses, improving overall brain function. Animal studies have demonstrated that RNA treatments boost neuroplasticity, leading to enhanced learning and memory. These therapies are being evaluated as a method to enhance cognitive abilities, potentially benefiting even healthy individuals.

Research indicates that specific cognitive domains respond differently to RNA-based interventions. Working memory and cognitive flexibility show the most substantial improvements in animal models, while processing speed and attention demonstrate moderate enhancements. Executive functions—particularly inhibitory control and planning—appear responsive to treatments targeting the prefrontal cortex. Episodic memory formation, a function heavily dependent on hippocampal activity, also shows promising responses to RNA therapy, particularly for age-related decline.

The mechanism involves targeting specific genes to enhance memory and fine-tune postsynaptic receptors, as demonstrated in preclinical studies. Initial research shows these RNA-based approaches can improve memory recall, although further studies on safety and toxicity are needed.

Multiple animal studies have confirmed these treatments' effectiveness, raising the possibility of future applications to augment human intelligence, even in healthy individuals. These results demonstrate the potential of gene therapies, with significant ramifications for human cognitive enhancement.

Cognitive Enhancement and Ethical Considerations

RNA-based therapies are emerging as a groundbreaking approach to treating mental health conditions such as anxiety, depression, and neurodegenerative diseases. These treatments offer significant advantages over conventional methods, providing long-term efficacy, personalized interventions, and minimal side effects, while reducing the frequency of dosing.

Compared to other enhancement approaches, RNA neuromodulators offer several distinct advantages. Unlike pharmacological enhancers (such as modafinil or methylphenidate) that broadly affect neurotransmitter systems, RNA therapies can target specific neural circuits with greater precision. This targeted approach potentially reduces side effects while producing more substantial and lasting improvements. In contrast to transcranial magnetic stimulation or direct current stimulation, which temporarily modify neural activity, RNA therapies can induce persistent changes in gene expression, potentially leading to durable cognitive enhancements. However, these advantages come with higher development costs and more complex delivery challenges than existing approaches.

For example, RNA therapies targeting overactive brain circuits have shown promise in alleviating anxiety and improving memory by modulating key receptors like 5-HT2a without notable adverse effects. As research progresses, RNA-based treatments are seen not only as a means of managing symptoms but as a way to address the root causes of many neuropsychiatric disorders, potentially slowing or halting disease progression.

However, the development of these advanced therapies also raises critical ethical concerns. One of the most significant debates in genetic engineering revolves around the ethics of human enhancement, particularly following a controversial experiment in China. In that experiment, CRISPR-Cas9 was used to edit the genes of three newborns to prevent HIV infection, sparking global outrage and prompting the scientific community to grapple with the ethical implications of germline editing. The concern primarily centers on the modification of germline cells, which can pass changes to future generations, posing risks of unintended consequences.

The regulatory landscape for RNA-based cognitive enhancement remains under development. Currently, these therapies would fall under existing frameworks for gene therapies and biologics, requiring rigorous clinical trials and safety demonstrations before approval. However, as applications expand beyond treating disorders to enhancing normal function, regulatory bodies like the FDA in the United States and the EMA in Europe are considering specific guidelines for "enhancement" versus "treatment" applications. These distinctions may influence approval pathways, insurance coverage, and accessibility of these technologies once they reach clinical application.

An important aspect of RNA-based cognitive enhancement technologies is that they target somatically differentiated cells, not germline cells, ensuring that any modifications remain within the treated individual and are not passed to offspring.

Future Directions

The future of RNA-based cognitive enhancement is promising. Advances in molecular engineering and nanotechnology are refining delivery mechanisms, making these therapies more effective and less invasive. Personalized medicine and combination therapies are also on the horizon, offering tailored treatment plans that address specific patient needs while enhancing overall outcomes.

These deelopments will likely redefine mental health care through targeted interventions that address root causes rather than merely managing symptoms. The potential to restore cognitive function in those with impairments while enhancing capabilities in healthy individuals represents a significant frontier in neuroscience—one that could fundamentally transform our understanding of human cognitive potential.

An important aspect of RNA-based cognitive enhancement technologies is that they target somatically differentiated cells, not germline cells, ensuring that any modifications remain within the treated individual and are not passed to offspring

The mechanism involves targeting specific genes to enhance memory and fine-tune postsynaptic receptors, as demonstrated in preclinical studies. Initial research shows these RNA-based approaches can improve memory recall, although further studies on safety and toxicity are needed. Multiple animal studies have confirmed these treatments' effectiveness, raising the possibility of future applications to augment human intelligence, even in healthy individuals. These results demonstrate the potential of gene therapies, with significant ramifications for human cognitive enhancement.

Cognitive Enhancement and Ethical Considerations

RNA-based therapies are emerging as a groundbreaking approach to treating mental health conditions such as anxiety, depression, and neurodegenerative diseases. These treatments offer significant advantages over conventional methods, providing long-term efficacy, personalized interventions, and minimal side effects, while reducing the frequency of dosing.

Compared to other enhancement approaches, RNA neuromodulators offer several distinct advantages. Unlike pharmacological enhancers (such as modafinil or methylphenidate) that broadly affect neurotransmitter systems, RNA therapies can target specific neural circuits with greater precision. This targeted approach potentially reduces side effects while producing more substantial and lasting improvements. In contrast to transcranial magnetic stimulation or direct current stimulation, which temporarily modify neural activity, RNA therapies can induce persistent changes in gene expression, potentially leading to durable cognitive enhancements. However, these advantages come with higher development costs and more complex delivery challenges than existing approaches.

For example, RNA therapies targeting overactive brain circuits have shown promise in alleviating anxiety and improving memory by modulating key receptors like 5-HT2a without notable adverse effects. As research progresses, RNA-based treatments are seen not only as a means of managing symptoms but as a way to address the root causes of many neuropsychiatric disorders, potentially slowing or halting disease progression.

However, the development of these advanced therapies also raises critical ethical concerns. One of the most significant debates in genetic engineering revolves around the ethics of human enhancement, particularly following a controversial experiment in China. In that experiment, CRISPR-Cas9 was used to edit the genes of three newborns to prevent HIV infection, sparking global outrage and prompting the scientific community to grapple with the ethical implications of germline editing. The concern primarily centers on the modification of germline cells, which can pass changes to future generations, posing risks of unintended consequences.

The regulatory landscape for RNA-based cognitive enhancement remains under development. Currently, these therapies would fall under existing frameworks for gene therapies and biologics, requiring rigorous clinical trials and safety demonstrations before approval. However, as applications expand beyond treating disorders to enhancing normal function, regulatory bodies like the FDA in the United States and the EMA in Europe are considering specific guidelines for "enhancement" versus "treatment" applications. These distinctions may influence approval pathways, insurance coverage, and accessibility of these technologies once they reach clinical application.

9. The Science of Stress Reduction

Chronic stress is a widespread issue globally, affecting a significant portion of the population. Studies show that up to seventy-five percent of adults report experiencing some level of stress in the previous month, though chronic stress affects a smaller but significant subset. The impacts of this pervasive condition extend far beyond momentary discomfort—they reach into the very structure and function of our brains.

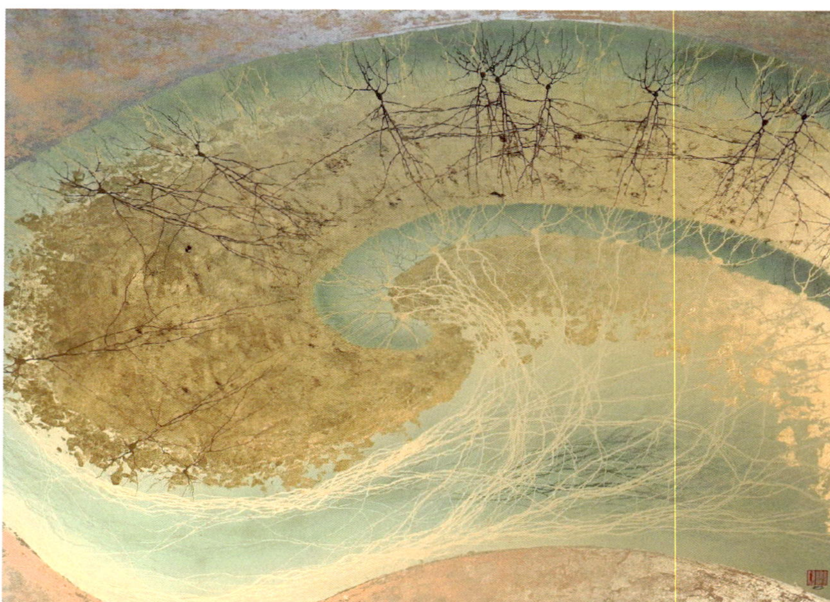

The Hippocampus, Greg Dunn

Chronic stress, as defined by the American Psychological Association, refers to long-term exposure to stress, and is linked to six primary factors contributing to severe health issues and death, including cardiovascular disease, cancer, and diabetes.

Scientists engaged in cutting-edge research are now investigating innovative treatments for chronic stress to reduce its impact on the brain and psychological well-being. Researchers seek to employ sophisticated RNA neuromodulators to precisely target and control gene expression. This strategy shows remarkable potential as a viable method for reducing the effects of chronic stress on the brain.

Biological Mechanisms Underlying Stress

Chronic stress disturbs the equilibrium of neurotransmitters, the chemical agents responsible for conveying messages between synapses in the brain. Scientists may now use RNA-based therapeutics in preclinical animal trials to control the release and modulation of neurotransmitters such as serotonin, dopamine, and norepinephrine, which are frequently disturbed by prolonged stress. This method aims to improve mood while also reducing anxiety and hypervigilance caused by long-term stress.

RNA neuromodulators may have the ability to selectively target and manipulate the production of serotonin, a vital neurotransmitter that plays a key role in regulating mood. Traditionally, selective serotonin reuptake inhibitors (SSRIs) are medicines suggested for relieving symptoms of depression and anxiety caused by prolonged stress, but these medications often come with unwanted side effects and limited efficacy for many patients.

Neuroinflammation and Alterations in the Structure and Function of the Brain

Leading scientists employ RNA interference (RNAi) to regulate gene expression by suppressing certain neuronal receptors that drive stress responses. RNAi-based treatments offer a precise and effective method for addressing the effects of chronic stress by targeting genes that produce hyperactivity and stress hormones.

MicroRNAs (miRNAs)

Scientists are investigating the potential of microRNAs to alter the pathways associated with inflammation in specific brain regions. By regulating the influence of microRNA, these treatments can restore brain function to a healthy condition and mitigate the negative effects of chronic stress. MicroRNAs are important for neuronal well-being. RNA treatments may reduce neuroinflammation and enhance the survival and functioning of neurons in chronic stress conditions.

Treatments Utilizing Messenger RNA (mRNA)

Scientists are creating messenger RNA (mRNA) and developing therapeutic proteins specifically tailored to mitigate the harmful effects of long-term stress. These new medications may have neuroprotective properties. Treatments like this will enable a more adaptable and accurate approach to treating stress-related illnesses.

Alleviating Chronic Stress

RNA neuromodulators present opportunities to address disorders that result from chronic stress. These medicines specifically target genes that produce, release, and activate neurotransmitter receptors. They help to restore a balance in the levels of neurotransmitters in the body.

Scientists are documenting how RNA therapeutics can rectify neurotransmitter imbalances and address and cognitive impairments resulting from prolonged stress by explicitly targeting receptors responsible for the imbalance.

These RNA treatments, addressed to modulate serotonin signaling, reduce anxiety and stress levels in preclinical studies and can alter expression of serotonin receptors, leading to lowering serotonin signaling and resulting in anxiety reduction. RNA neuromodulators have the potential to modify specific brain circuits, increasing emotional well-being and motivation in persons dealing with chronic stress.

In early stages of neurodegeneragation RNA treatments in early stages of development, may enhance the stimulation of genes linked to synapse formation and neuronal development, augmenting efficacy in neuroplasticity. The objective is to improve an individual's ability to manage stress and facilitate a recovery process from the brain damage resulting from prolonged stress.

RNA-based therapies can enhance neuroplasticity by boosting the activation of genes that play a role in the creation of synapses and the growth of neurons. This approach would aim to optimize an individual's capacity to manage stress and expedite the recovery of brain impairments resulting from stress. Additionally, administering RNA therapies to increase miR-132 levels may improve development of new synapses. Scientists are investigating potential of using microRNAs to modify the pathways associated with inflammation in brain areas.

Synaptic Plasticity

The phrase "synaptic plasticity" refers to synapses' ability to become stronger or weaker over time in response to increased or decreased activity. This flexibility is critical to learning, memory formation, and the brain's general adaptability. The brain's capacity to facilitate modifications in the structure and function of synapses enables it to adjust to novel information.

Dendrites, which are intricate extensions of neurons, have the role of receiving impulses from neighboring neurons. The relationship between miR-132 and increased dendritic development and complexity enhances the neurons' capacity to establish connections and facilitate efficient communication. microRNA-132 can alter the expression of essential target genes, leading to an enhancement in synaptic strength. This enhancement is necessary for memory consolidation and reconfiguring neural circuits responding to stimuli.

Dendritic Growth

RNA-based therapeutics aim to specifically target particular molecules of RNA in the body to enhance or inhibit their function. Particularly relevant are those RNA molecules that act on neurons and their morphological shape and structure. Among the various structures that characterize a neuron, there are "spines" distributed across their branched extension, called "dendrites," that receive electrical impulses from neighboring neurons.

Dendritic arborization is the growth and branching of dendrites. The cross-contact process of neuronal "talks" via their dendrites sustains the forming and maintaining of complex neural networks for brain function. The structural framework that scaffolds for dendrites is the cytoskeleton. The neuronal cytoskeleton supplies dendrites with structural integrity, enabling them to keep their shape, support their function, and demonstrate flexibility by expanding and generating new branches.

Since miR-132 regulates the proteins that govern the cytoskeleton, increasing the expression of miR-132 in neurons may lead to an elevation of miR-132 levels inside neurons. These treatments can improve the ability of neurons to change and adapt, leading to the creation of new connections between them.

Implications for Neurological Disorders

Improvements in dendritic development and synaptic plasticity may significantly impact therapy for neurological diseases. The interaction between these two processes generates this potential. Several types of cognitive decline, such as Alzheimer's and Parkinson's disease, are characterized by the reduction of synapses and complexity of dendrites. RNA-based therapies may restore synaptic connections. This may result in improved cognitive functions in individuals beset with these conditions. As highlighted before, one approach may be to elevate the intra-neuronal levels of miR-132.

Clinical Trials and Case Studies: Long-Term Safety and Efficacy

Leading-edge researchers are at the forefront of studies investigating the potential of RNA therapies in treating stress-related disorders. Pre-clinical and clinical studies are evaluating the safety and efficacy of different RNA-based treatments, with siRNAs, miRNAs, and mRNA therapies. Early results indicate interventions have the potential to enhance adaptability to stress and reduce the manifestation of stress-related symptoms.

Real-world studies demonstrate the potential efficacy of RNA therapies. RNA interference (RNAi) therapies that specifically target pro-inflammatory cytokines have shown a substantial reduction in neuroinflammation and improvement in cognitive function in animal models. However, these treatments still need to undergo clinical evaluation in humans.

Extensive studies are essential to confirm the long-term benefits and safety of these treatments. Preclinical studies, and human trials where feasible, help determine the optimal dosage, tolerability, and duration of therapeutic effects, while also monitoring for potential delayed adverse reactions.

Predictive Biomarkers for Chronic Nervous System Disorders

Leading researchers are advancing RNA therapeutics to address chronic stress and its neuropsychiatric effects. RNA-based drugs have the potential to offer highly effective and personalized therapy by explicitly targeting the essential biochemical processes involved in stress responses. Symptoms of chronic stress may be relieved at their root cause by using these methods and pave the way for significant advancements in managing this prevalent disorder.

By employing distinct lifespan indicators, researchers gain new insights into the long-term impact of chronic stress on brain health and aging. Current initiatives focus on unraveling hippocampal hyperactivity, which is increasingly recognized as a contributing factor in various neuropsychiatric and neurodegenerative disorders, including Alzheimer's disease, schizophrenia, and epilepsy. Recent research is exploring if and how RNA neuromodulators can be a viable method for reducing neuronal hyperactivity in the hippocampus. Ideally, RNA therapeutics are expected to minimize acute stress symptoms while postponing cognitive decline and other neurological problems.

Scientists from several fields are now conducting trials to investigate the potential use of RNA therapeutics to treat stress-related diseases. There is an urgent need for continued research in this specific field, particularly to increase our knowledge about the neurobiological bases of hippocampus hyperactivity.

Hyperactivity in the hippocampus has been associated with a variety of neuropsychiatric and neurodegenerative illnesses, including Alzheimer's, schizophrenia, and epilepsy. These disorders over time are linked to cognitive impairment. Researchers are investigating the possibility that exerting control over this hyperactivity through RNA treatments might ease the acute symptoms of stress while also slowing mental decline and other neurological changes associated with aging.

10. Memory Decline and Cognitive Impairment

Cognitive decline and memory loss are serious public health issues that require our attention. These chronic conditions can occur over time, with significant social and economic consequences. Result? Broken communities, higher hospitalization rates, and escalating healthcare costs.

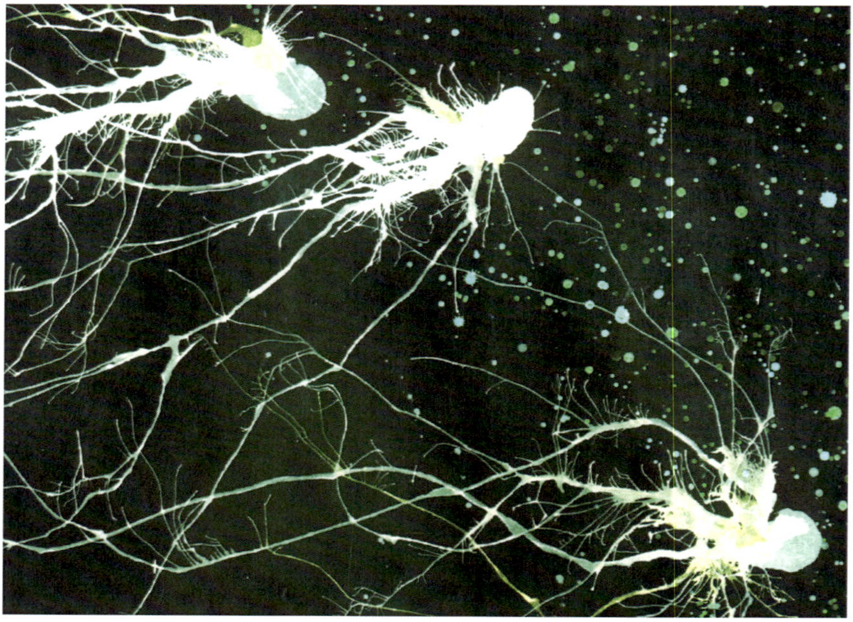

Purkinje Neurons, Greg Dunn

As of 2023, the United States has incurred an economic cost of approximately $105 billion due to disorders that result in cognitive deterioration. These illnesses place a growing financial strain on society—a burden that intensifies as the expense of treating particular conditions increases. Due to rising average lifespans and the growing elderly population, these costs are estimated to exceed one trillion dollars annually by 2050.

Classification of Affected Memory Types

Memory and cognitive deficits affect several forms of memory, each leading to distinct consequences. To understand these impacts, it helps to have a basic understanding of different memory types. While memory classification is complex and multifaceted (for which an interested reader can consult a neurology textbook such as Kandel et al.'s "Principles of Neural Science," 2021), certain memory systems are particularly relevant to our discussion of neurodegenerative and memory disorders:

Episodic memory refers to the ability to remember specific events and experiences, including the detailed characteristics of their surrounding circumstances. Cognitive decline in this area is often the first sign of Mild Cognitive Impairment (MCI) and Alzheimer's disease (AD). Severe decline in episodic memory significantly impairs an individual's capacity to remember prior events, negatively affecting their daily functioning and sense of self.

Semantic memory refers to the comprehensive understanding and acquired knowledge of the world, which includes both factual information and conceptual frameworks.

Working memory is a cognitive function necessary for rational thinking, decision-making, and behavior control. It involves the temporary storage and manipulation of information. Deterioration in this domain can impact an individual's capacity to perform complex activities, strategize, and engage in multitasking, resulting in heightened reliance on others and a diminished standard of living.

Procedural memory pertains to the recall of how to perform actions and skills. Interestingly, this form of memory is often preserved in neurodegenerative diseases until advanced stages. Activities such as riding a bicycle or typing rely on procedural memory. When this memory system eventually deteriorates, it can result in a devastating loss of independence.

Wider Ramifications

Memory and cognitive impairments extend far beyond difficulties recalling names or dates; they lead to significant problems in social, occupational, and physical aspects of life. These disorders are characterized by cognitive deficits, including diminished memory, attention, and processing speed, which impede daily functioning and diminish overall quality of life.

Approximately forty percent of individuals aged 65 and above experience some form of cognitive decline. This statistic indicates the extensive prevalence of these conditions and highlights an urgent requirement for effective treatments—representing a significant unmet need in human healthcare.

The Complexity of Neurocognitive Disorders

Neurocognitive disorders include a range of conditions, spanning moderate cognitive impairment to severe dementia. These illnesses impact more than 50 million individuals worldwide, with over 20 million people affected in the United States alone. They significantly influence the number of years spent living with disability.

There is an essential connection between the duration of living with a disability and the emergence of these particular diseases. They tend to coexist with other medical conditions, such as metabolic disorders and systemic diseases like hypertension. Over time, patients with neurodegenerative disorders experience decreased mobility and increased vulnerability to trauma. The transition from MCI to more severe illnesses such as Alzheimer's disease is marked by deteriorating symptoms and escalating healthcare requirements. This decline ultimately leads to higher mortality rates in the latter stages of illness.

Underlying Physiological Mechanisms

The accumulation of evidence and recent research has deepened our understanding of the intricacies associated with neurocognitive illnesses, highlighting the need to consider several pivotal elements. An important new direction, is directly modulating neuronal receptors in the limbic area in the brain associated with neurocognitive impairments. The abnormal neuronal activity found in conditions such as Alzheimer's disease, epilepsy, and schizophrenia disrupts the usual balance between excitatory and inhibitory signals in the brain.

An imbalance of this kind can lead to cognitive impairments, such as memory deficits and challenges in spatial orientation. During the early stages of Alzheimer's disease, increased metabolic activity in the hippocampus occurs before any noticeable decline in cognitive function. This suggests that targeting the reduction of excessive activity in the hippocampus might be a direct strategy for preventing and managing the illness.

Current treatment options primarily consist of SSRIs and other pharmaceuticals developed decades ago. Frequently, however, they do not provide sufficient relief and are associated with significant adverse effects. The limitations of these traditional treatments highlight the urgent need for novel therapeutic approaches.

Using the novel RNA-based therapeutic approaches discussed in previous chapters, we can engineer precisely targeted medications that utilize RNA interference (RNAi). Short hairpin RNA (shRNA) can silence specific genes, offering an accurate approach for precision genetic modification. RNA neuromodulators can modulate neurotransmitter systems that play a role in cognitive and emotional regulation without the adverse side effects commonly associated with conventional drugs.

Creating and implementing targeted RNA therapies to treat memory decline and cognitive impairments is urgent. Progress in genetic neuroengineering is improving the effectiveness of treatments that address the root causes of these diseases. When these treatments become available, we may see significant improvements in the well-being of many affected individuals.

Genetic and Molecular Pathways

The involvement of biochemical pathways, especially those controlled by 5-HT2a receptors, is highlighted in relation to these disorders. Overstimulation of these receptors exacerbates anxiety symptoms and hampers cognitive functions. Genetic treatments targeting these pathways, such as using shRNA to control receptor activation, have shown promise in reducing symptoms and restoring normal cognitive functioning.

Excessive activity in deep brain circuits, particularly in the hippocampus, results in significant network disruptions that can cause cellular damage. This damage is a crucial factor in worsening conditions like Alzheimer's, highlighting the need for therapeutic methods that can provide early diagnosis and intervention, preserving normal and healthy cerebral function.

Therapeutic Implications and Innovations

The 5-HT2a post-synaptic receptors play a crucial role in these circuits. Excessive stimulation of these receptors leads to worsening anxiety symptoms and impaired cognitive processes, resulting in several adverse outcomes.

We must devise novel treatment approaches to address the intricacies of neurocognitive illnesses. Genetic neuroengineering techniques are now widespread in research. They offer the precise capability to target and modify specific brain circuits. These genetic tools can be administered noninvasively and efficiently transported to the brain. These compounds then regulate gene expression, providing a non-intrusive and precise therapeutic approach.

Research on treatment improvements reveals the intricate and varied features of neuropsychiatric and neurocognitive disorders and underscores the importance of a comprehensive and interdisciplinary approach to their treatment. The convergence of advanced genetic technologies and improved understanding of brain function (and dysfunction) offers new possibilities for effective medications that address both the symptoms and underlying causes of these complex illnesses.

RNA neuromodulators hold particular promise in this area, as they can target the specific neural mechanisms involved in memory formation and cognitive processing with unprecedented precision. As this field advances, we may soon witness a fundamental shift in how we approach the treatment of memory and cognitive disorders—moving from symptom management to addressing the root biological causes of these devastating conditions.

11. Neurocognitive Stress Cascade (NSC): RNA Neuromodulator Treatment

Overview of the Stress Cascade

The stress cascade represents a complex interplay of cellular neurochemical and physiological processes evolved to mobilize the body's resources against perceived threats. It involves the coordination of multiple brain regions, hormones, and feedback loops. We're continuing to learn more, with findings shedding light on how chronic stress transitions into dysfunction and how RNA-based neuromodulators may effectively intervene in this process (Sarkar et al., 2021).

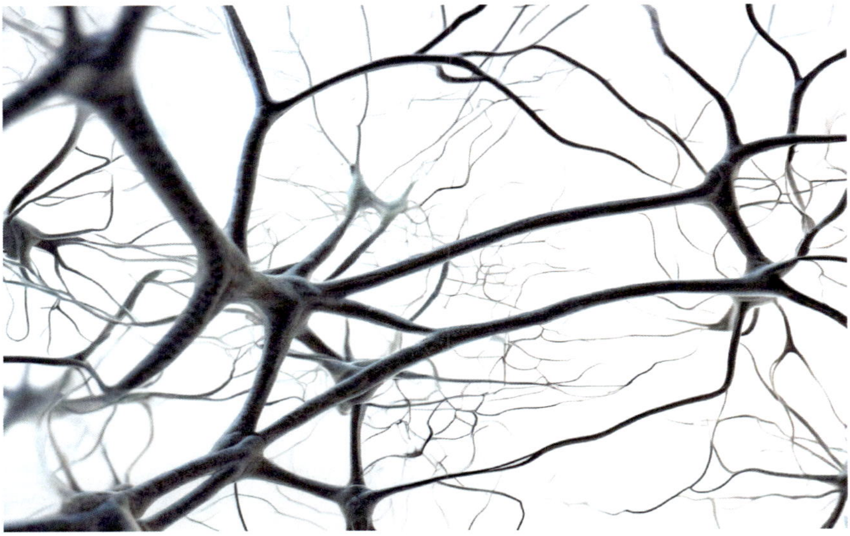

Pyramidal Neurons

Chronic Stress and Its Impact

Prolonged HPA Axis Activation: Chronic stress transitions from an acute response to sustained activation of the hypothalamic-pituitary-adrenal (HPA) axis, resulting in elevated cortisol levels.

Neuronal Damage in the Hippocampus: Persistently high cortisol levels contribute to neuronal damage, particularly in the hippocampus—a brain region critical for memory consolidation and emotional regulation. This aligns with well-established stress-related pathophysiology.

Therapeutic Insights from Preclinical Studies

RNA neuromodulators have demonstrated promise in mitigating the detrimental effects of chronic stress (Morrison et al., 2021):

Reduction in Neuronal Hyperactivity: RNA therapies effectively decreased hyperactivity in key brain regions such as the hippocampus and amygdala.

Normalization of Cortisol Levels: Treatment restored HPA axis balance, reducing cortisol levels and improving stress responses.

Behavioral Resilience: Rodent models showed significant improvements in stress-related behaviors, underscoring the therapeutic potential of RNA-based interventions to enhance resilience and cognitive recovery.

Perception and Initial Brain Response

1. *Hypothalamus Activation:* Serves as the central command for initiating the stress response. Engages the sympathetic nervous system and HPA axis to coordinate physiological adaptations to stress.
2. *Amygdala Activation:* The amygdala acts as the brain's alarm system, identifying and processing threats. It signals the hypothalamus to initiate the "fight or flight" response, preparing the body to respond to perceived danger.
3. *Sympathetic Nervous System Activation:* The hypothalamus activates the sympathetic nervous system, resulting in rapid physiological adjustments, such as increased heart rate, heightened alertness, and energy mobilization. These changes enable the body to confront or evade the threat effectively.

The integrated approach of the stress cascade highlights potential intervention points where RNA neuromodulators could mitigate the progression of chronic stress, restore homeostasis, and enhance both emotional and cognitive well-being (Yang et al., 2021). The accumulating insights lay a strong foundation for advancing clinical research and therapeutic innovations.

Chronic Stress and the HPA Axis

Prolonged activation of the HPA axis drives sustained stress responses, contributing to systemic physiological and psychological strain. Chronic stress results in the prolonged activation of the hypothalamic-pituitary-adrenal (HPA) axis, the body's central stress-regulation system. In response to sustained stress, the hypothalamus releases corticotropin-releasing hormone (CRH), which signals the pituitary gland to produce adrenocorticotropic hormone (ACTH). This, in turn, stimulates the adrenal glands to release cortisol, the primary stress hormone.

Cortisol's Dual Role: Essential for Survival, Harmful Over Time

Cortisol is a critical hormone for maintaining the stress response, providing immediate energy by increasing glucose availability, suppressing non-essential functions like digestion, reproduction, and immune activity, and prioritizing resources to ensure survival.

Typically, cortisol levels are tightly regulated through feedback mechanisms. However, chronic stress leads to prolonged cortisol elevation, resulting in significant health risks. Persistent immune suppression increases susceptibility to infections and slows healing processes. Metabolic disruptions, including weight gain concentrated in the abdominal area, are common.

Most concerning is the impact on the hippocampus in the limbic area—a brain region essential for memory and learning—where high cortisol levels lead to conditions that damage neurons, leading to cognitive impairments. This combination of physiological impacts on the brain are profound and widespread. There are downstream effects of chronic stress on overall health and well-being.

Feedback Mechanisms and Shutdown

Calcium Dysregulation and Cellular Excitotoxicity

1. Disrupted Calcium Balance: Chronic stress disrupts calcium homeostasis in neurons, leading to excitotoxicity—a condition characterized by excessive calcium influx.
2. Destructive Enzyme Activation: High intracellular calcium levels activate harmful enzymes that degrade critical cellular components, including mitochondria, the cell's energy producers.
3. Mitochondrial Dysfunction and Cell Death: Prolonged mitochondrial stress leads to dysfunction and triggers apoptosis (programmed cell death). Over time, this progressive loss of neurons undermines brain function, contributing to cognitive decline (Kumar et al., 2021).

Plaque Accumulation and Neuronal Degeneration

Toxic Protein Release: Excitotoxicity stimulates the production of toxic proteins, such as amyloid-beta (Aβ) and phosphorylated tau. These proteins aggregate into plaques and neurofibrillary tangles, perpetuating neuronal damage and synaptic loss.

Worsening Calcium Dysregulation: Amyloid-beta plaques exacerbate calcium dysregulation, intensifying cellular stress and further accelerating neuronal death.

Impact on Cognitive Function: This pathological cascade particularly affects the hippocampus, a brain region essential for memory and learning. As hippocampal neurons degenerate, memory loss and cognitive disruptions become more pronounced.

The Cumulative Impact of Calcium Dysregulation, Excitotoxicity, and Plaque Formation

The interplay of calcium dysregulation, excitotoxicity, and plaque formation profoundly impairs the hippocampus, a brain region essential for memory formation and spatial navigation. Damage to the hippocampus results in memory deficits, cognitive decline, and a progressive loss of executive functions. These effects are hallmarks of both early-stage Alzheimer's disease (AD) and chronic anxiety-related cognitive impairments.

The Role of Limbic Circuit Dysfunction in Cognitive Decline

This cascade of neural damage underscores the critical role of limbic circuit dysfunction in cognitive decline, particularly in memory processing and executive function. The associated conditions and symptoms often include:

- Memory deficits
- Cognitive decline
- Loss of executive functions
- Anxiety-related impairments
- Calcium dysregulation
- Impaired spatial navigation
- Plaque formation

Potential Treatment Strategies for Alzheimer's Disease Stages with RNA Therapy

As a continuation of this theoretical framework outlining the progression of Alzheimer's disease, associated symptoms, and potential interventions at each stage, we'll look at RNA-based targeted therapies in conjunction with supportive care and medications aimed at improving quality of life.

Potential Treatment Strategies with RNA Neuromodulators for Stages of Alzheimer's Disease

1. Stage: Pre-Diagnosis

Symptoms: Forgetfulness, mild anxiety

Treatment: Monitor symptoms with a "watch and wait" approach. Provide supportive care to manage mild cognitive and emotional concerns. Consider prescribing SSRIs or benzodiazepines as needed for anxiety management.

2. Stage: Early-onset Alzheimer's Disease (EOAD)

Symptoms: Worsening memory loss, early cognitive decline.

Treatment: Begin targeted RNA therapy for reducing limbic circuit inflammation, to address cognitive and psychiatric symptoms (Kumar et al., 2021).

3. Stage: EOAD with Cognitive Issues

Symptoms: Continued memory loss, diminishing cognitive abilities, and increasing difficulty with daily tasks.

Treatment: Intensify therapeutic approach with a combination of targeted RNA therapy for reducing limbic circuit inflammation to stabilize cognitive function and mitigate further decline (Morrison et al., 2021).

4. Stage: Moderate to Severe Disease

Symptoms: Progression to advanced memory and cognitive impairment, with pronounced symptoms affecting daily living.

Treatment: Integrate amyloid-targeting drugs alongside targeted RNA therapy to reduce limbic circuit inflammation.

5. Stage: Advanced/Severe Disease

Symptoms: Severe cognitive impairment has a significant impact on quality of life across all measures.

Treatment: Focus on quality-of-life medications to alleviate discomfort and enhance well-being. RNA therapy for reducing limbic circuit inflammation to stabilize cognitive function. Continue supportive care to ensure comfort and dignity in this stage.

Administering RNA Therapy for Reducing Limbic Circuit Inflammation to Stabilize Cognitive and Psychiatric Functions

As further research and clinical trials are complete, we may have a range of genetic medicines for the treatment of anxiety, memory impairments, and related cognitive challenges by targeting specific neurons in specific neural circuits (Bhat et al., 2022). Imagine a future where anxiety and cognitive decline are treated at their cellular biomechanical site, with minimal side effects and long-term benefits. A new future opens up with the one-time administration of a short hairpin RNA (shRNA)-based therapeutic delivered intranasally to modulate the brain's stress and memory centers.

Immediate Impact: Modulating the Stress Response

The transformation journey starts immediately. The RNA therapeutic bypasses the cribriform plate and enters the brain system. Theoretically, upon delivery, the RNA knocks down hyperactive 5-HT2a receptor activity in critical regions such as the hypothalamus and amygdala (Taliaferro et al., 2016). This first step lowers the release of corticotropin-releasing hormone (CRH), effectively reducing cortisol levels and calming overactivation of the hypothalamic-pituitary-adrenal (HPA) axis. The amygdala's response to stress diminishes, quieting the overwhelming anxiety triggers that have previously hijacked emotional stability.

Restoring Cognitive Balance

If results in humans follow the preclinical pattern, as cortisol production normalizes, the brain's natural feedback mechanisms regain control. Overactive neurons in the hippocampus recover their natural state, enabling clearer memory retention and cognitive processing. Synaptic plasticity improves, fostering long-term potentiation (LTP), the cellular foundation of learning and memory. For individual patients, this translates to sharper thinking, better emotional regulation, and an enhanced ability to cope with stress.

A Targeted and Lasting Solution

Unlike traditional drugs like diazepam or SSRIs, which broadly sedate the brain or mask symptoms, an RNA therapy can achieve this with pinpoint precision. By targeting the molecular pathways driving stress and hyperactivity, these therapies may offer sustained benefits without sedation, dependency, or the systemic side effects typical of older treatments (Wu & Kapfhammer, 2021).

Proven Efficacy in Preclinical Studies

RNA neuromodulators have demonstrated remarkable efficacy in laboratory studies. Neuroimaging confirmed targeted delivery and receptor modulation. Behavioral tests, such as the light/dark box test, showed anxiety reductions of up to ~40%, comparable to diazepam, but without its sedative drawbacks. Memory improvements reached over 100% in recognition tasks, highlighting the dual benefit of addressing emotional and cognitive challenges. Even after weeks, these effects persisted, underscoring the long-term stability these therapies can provide.

Innovative Delivery for a New Era of Treatment

The intranasal delivery system, leveraging adeno-associated viruses (AAVs), represents a technological leap (Yang et al., 2021). This method ensures that therapeutic agents bypass the blood-brain barrier, directly accessing affected brain regions. It's a far cry from the blunt tools of the past, embodying a precise, personalized approach to mental health treatment.

We envision a future where treatments don't just suppress symptoms but repair the neural imbalances directly at the neuron level of mental health conditions (Addis et al., 2011). By addressing the biomechanics of anxiety and cognitive dysfunction, these therapies are leading a paradigm shift, offering not just relief but recovery, resilience, and a return to emotional and mental well-being. This is not just innovation—it's transformation.

Administration of RNA Therapeutic and Comparative Efficacy with Established Medications

Mechanism of Action

shRNA compounds may specifically reduce hyperactivity in the amygdala and hippocampus, key regions involved in stress processing and memory (Wu & Kapfhammer, 2021). By impacting the HPA axis, the RNA therapeutic can reduce excessive cortisol production. The body starts transitioning from a heightened state of alert to a baseline. The parasympathetic system gradually regains dominance, aiding in returning heart rate, blood pressure, and other immediate stress responses to normal.

Neuroprotective Effects

With sustained, balanced activity, the RNA therapeutic helps protect against long-term damage, such as inflammation and cell death (Kumar et al., 2021).

Sustained Cognitive Optimization

Continued downregulation prevents relapses of chronic stress responses, supporting ongoing cognitive health and balance. It targets HPA axis feedback by lowering cortisol levels, allowing the body's natural feedback loop to recalibrate for a more sustainable balance in stress response without reliance on external sedative effects. It promotes synaptic plasticity and neuroprotection in the hippocampus, facilitating recovery from stress-induced damage and supporting memory retention. Sustained cognitive and emotional stability is achieved by rebalancing stress pathways, reducing relapse into chronic stress states, and enhancing resilience through targeted genetic modulation (Sarkar et al., 2021).

Diazepam Comparison

Taken for acute anxiety, it rapidly increases GABA activity and produces a calming effect but affects the entire brain and body, leading to sedation and dependency. Diazepam potentiates GABAergic inhibition, resulting in generalized sedation that reduces anxiety symptoms but lacks neurochemical specificity, which can impair cognitive functions and motor skills. Diazepam provides temporary relief of symptoms without restoring HPA axis or feedback mechanisms, often requiring repeated doses, which disrupt natural stress responses and may lead to dependency.

Diazepam offers no support for neuronal recovery or synaptic plasticity. Prolonged use can negatively affect neuroplasticity and cognitive health, especially with dependence and withdrawal effects. Diazepam offers temporary relief without promoting long-term stability. Continued use can lead to emotional blunting and dependency, with cognitive side effects worsening over time.

Valium Comparison

Similar to diazepam, Valium enhances GABA neurotransmission across the brain. It is effective for acute anxiety relief but lacks specificity, often causing drowsiness, tolerance, and withdrawal symptoms. It functions similarly to diazepam, broadly enhancing GABAergic signaling, which calms anxiety but may impair alertness, coordination, and cognitive clarity due to its non-selective action.

It does not restore feedback loops; it suppresses symptoms without addressing underlying dysregulation in cortisol or HPA signaling, leading to dependency with long-term use. Similarly, it does not support neuronal recovery. Its effects on cognitive rebalancing are limited to short-term anxiety reduction, often impairing cognitive flexibility in long-term users. It has similar limitations as diazepam, as it does not provide a path toward emotional or cognitive resilience and may contribute to worsening stability through tolerance and dependency.

Donepezil Comparison

Primarily used in Alzheimer's, administered orally to enhance acetylcholine levels, indirectly supporting cognition. Does not address acute stress or anxiety. Donepezil increases acetylcholine availability to enhance cognition but has a limited immediate impact on stress-related neurochemical pathways. It does not engage directly with stress response pathways, focusing instead on maintaining acetylcholine levels for memory. It does not impact the HPA axis or feedback mechanisms.

Donepezil supports cognitive function, slowing acetylcholine breakdown, providing stabilization for memory but limited recovery or neuroprotection in stress-related neuronal damage contexts. It provides mild cognitive support for dementia patients but does not explicitly address stress or emotional stability.

There is a clear and urgent need for innovative solutions that can address these limitations while also opening doors to broader applications in brain health.

The Opportunity

Advances in gene therapy and precision editing now make it possible to deliver targeted, noninvasive treatments directly to the brain (Wu et al., 2021). These days we see not just MCI and anxiety but a wide spectrum of neurological and psychiatric disorders, including cognitive decline, and anxiety being considered as indications.

Breakthrough RNA-Based Therapy

Researchers have developed a high-precision RNA editing technology that selectively targets the 5-HT2a neuronal receptor in the brain with no observed off-target effects (Bhat et al., 2022). It achieves noninvasive delivery via intranasal administration, making the treatment practical and scalable.

Demonstrated Efficacy

Preclinical studies in mice and rats show significant memory improvements and reduced anxiety. Comparable to psychedelic therapies but without the hallucinogenic side effects, enhancing safety and acceptability.

Safety and Scalability

Extensive testing reveals no toxicological concerns in preclinical. Leveraging RNA editing offers a temporary yet effective therapeutic approach, sidestepping the regulatory hurdles associated with permanent genetic modifications.

Versatility

The technology is not limited to MCI and anxiety; it lays the groundwork for addressing a broad array of brain health challenges, making this platform adaptable for future innovations.

Why It Matters

Precision gene therapy for the brain is poised to revolutionize how we treat neurological and psychiatric conditions (Kumar et al., 2021). By targeting specific mechanisms with unparalleled accuracy, RNA neuromodulator technology could address long-standing limitations of traditional pharmaceuticals, offering better outcomes and improved quality of life. The next steps will move these potential treatments from preclinical success to clinical trials to prove efficacy in humans. This platform technology has the potential to deliver breakthrough therapies for millions of patients while creating a foundation for sustained growth and impact in the expanding market of neurological and psychiatric treatments.

RNA neuromodulators intervene by downregulating hyperactivity in the amygdala and hippocampus, centers for stress processing, restoring balance to the HPA axis and promoting cognitive recovery (Morrison et al., 2021). Through this targeted modulation, RNA therapy helps re-establish homeostasis, reducing cortisol levels, enhancing resilience, and supporting long-term emotional stability.

12. Next-Generation Personal Change

By J. L. Mee

Combining cognitive neuro-engineering technology with human potential development techniques can help mentally well people discover newfound cognitive assets and leverage them into meaningful personal growth to achieve greater success and realize their dreams.

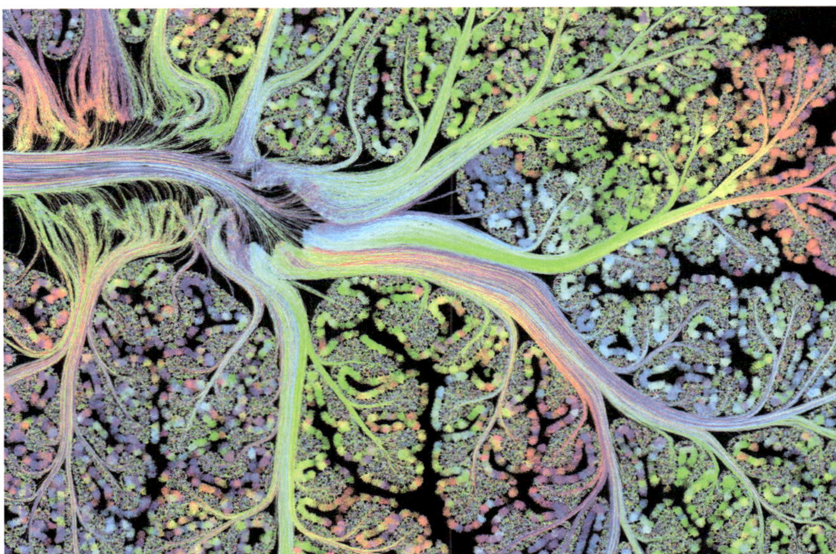

Cerebellar-Folia, Greg Dunn

Bridging a Gap Between Desire and Achievement

Emerging RNA cognitive therapies operate by mitigating the psychological barriers often encountered when an individual attempts to step beyond their comfort zone. This stepping out is essential for the adoption of new behaviors, a critical component of personal growth and transformation.

The primary hindrance in this process is the overwhelming feelings of stress and anxiety that accompany the venture into unfamiliar territory. RNA therapies, through their action, alleviate these feelings for a period of three to four months. This duration is strategically significant as it aligns with the time frame necessary for the formation of new habits and the crystallization of these behaviors into the individual's lifestyle. By doing so, it not only aids in the initial adoption of these behaviors but also ensures that they are sufficiently ingrained to persist as the new default mode of operation.

Supercharging the Formula for Change

Human potential development techniques such as neuro-linguistic programming and cognitive behavioral psychology emphasize the importance of overcoming past conditioning and rewiring the brain to form new, more empowering beliefs and habits. RNA cognitive therapies serve as a catalyst in this process, supercharging the formula by removing the "monkey mind"—the incessant, often negative internal dialogue that hinders focus and progress.

With RNA cognitive therapy, the typically dominant barriers of fear, doubt, and procrastination are significantly reduced, enabling individuals to fully engage with personal transformation practices. This includes deep meditation, visualization of future possibilities, and the embodiment of new states of being. By facilitating a clearer path for mental reprogramming, RNA cognitive therapy makes the once daunting task of significant personal change faster and more achievable.

Dramatic Results, Expanded Possibilities

The introduction of RNA cognitive therapy into the process of personal transformation can dramatically expedite the journey of change. Traditional methods of personal development often require years of dedicated practice, introspection, and gradual reconditioning of the mind. However, with the aid of RNA cognitive therapy, what used to take years can now be condensed into a matter of weeks or months.

An accelerated transformation like this opens up possibilities previously considered beyond reach for many individuals. Goals and aspirations that were once shelved as unrealistic or unattainable can now be actively pursued and achieved. Moreover, the quieting of the "monkey mind" for a significant period allows for a deeper and more effective rewiring of the brain.

Rewiring is not just superficial or temporary; it is a profound reconfiguration of thought patterns, belief systems, and behavioral responses. By the time the usual mental barriers begin to reassert themselves, the individual has already established a new set of neural pathways, supporting the new behaviors and thought processes. This effectively means that the individual has not only adopted new habits but has also transformed their identity to align with these new patterns.

Ushering in a New Era of Personal Development

RNA cognitive therapy represents a significant leap forward in the field of personal development. By addressing one of the most significant challenges in personal transformation—the ability to overcome the innate resistance to change—it opens up new horizons for individual growth and self-improvement. This next-generation approach offers a promising avenue for those seeking to enhance their lives, achieve their full potential, and realize their dreams in a fraction of the time previously thought necessary. RNA cognitive therapy is not just a tool for change; it is a gateway to a new realm of possibilities in the journey of personal evolution.

13. Cognitive Resilience Continuum (CRC)

The Cognitive Resilience Continuum (CRC) is a theoretical model to illustrate a spectrum of cognitive resilience, that spans from baseline functioning to enhanced, or optimized cognitive states. This framework emphasizes the dynamic potential to manage, adapt, and improve cognitive abilities over time, aligning with both therapeutic and optimization goals.

Cognition Continuum

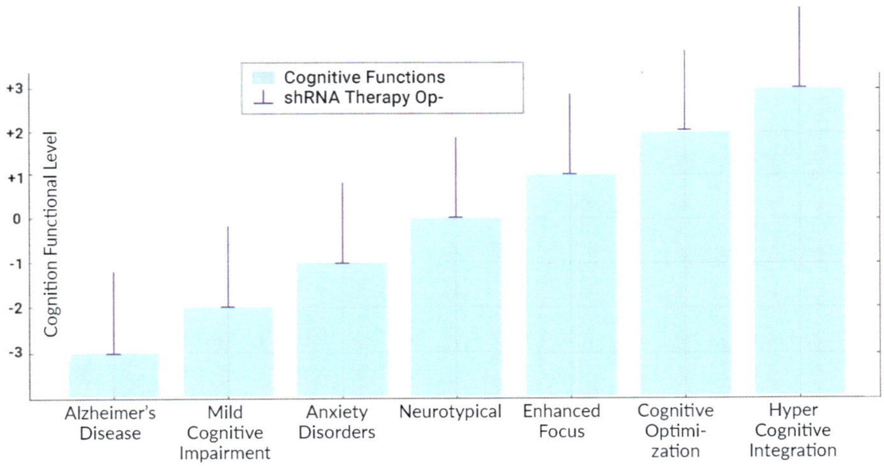

Cognitive Resiliance Continuum a Conceptual Model shRNA Therapy, Optimizing Cognitive Functions Across Spectrum of Cognitive States.

The Y-axis of the model illustrates cognitive resilience as a measure of adaptive improvement, charting a range from impaired states—such as Mild Cognitive Impairment (MCI) or Alzheimer's Disease (AD)—to optimized states that reflect cognitive optimization and resilience.

Cognitive States & Potential Cognitive Gains

Cognitive functioning is depicted along a continuum of measurable and durable states, offering a structured view of potential cognitive gains. Light blue bars within the model indicate baseline cognitive function for each state, providing a clear reference point for progression.

Zero-to-Plus-Three spans Enhanced Focus, Cognitive Optimization and Hyper-Cognitive Integration, describing a progression of refinement of cognitive skills to deeply integrated, peak-level mental functioning. The underlying neurobiological mechanisms in each stage illustrate how connectivity, neurotransmitter balance, and network synchrony contribute to increasingly complex cognitive capabilities.

Zero-to-Minus-Three reflect progressively worsening stages and neurobiological markers. From neurotransmitter imbalances in anxiety to structural degeneration in AD, with cumulative changes in brain structure, function, and chemistry leading to cognitive decline. A deeper look follows.

Parameters of Cognitive States

[+1] Enhanced Focus

Enhanced Focus characterizes individuals with improved attention, memory retention, and problem-solving abilities beyond the neurotypical range. These individuals can sustain attention and process information with greater accuracy and speed.

This state often results from cognitive training, healthy lifestyle practices, and genetic predispositions that favor optimal neurotransmitter balance and resilience against mental fatigue.

A balanced dopamine release and acetylcholine enhance focus, boosting attention and task-related motivation. Neurotransmitters enable fast synaptic responses in the prefrontal cortex and parietal regions, optimizing executive function and working memory.

[+2] Cognitive Optimization

Represents a heightened state of cognitive abilities in which individuals demonstrate advanced memory, strategic thinking, and creativity. They experience elevated resilience to mental fatigue and demonstrate incredible learning speed and adaptability.

Administer Genetic Neuromodulator

Combine with rigorous cognitive stimulation and lifestyle choices to enhance neuroplasticity and brain health. Individuals may also benefit from nootropic compounds or therapies to support and integrate neurogenesis and synaptic health.

Enhanced Synaptic Efficiency

High synaptic density in the prefrontal cortex, parietal cortex, and hippocampus enables rapid signal transmission, supporting complex decision-making and adaptive responses.

Optimized Neurotransmitter Balance

Higher levels of dopamine and acetylcholine sustain focus and motivation. Dopamine supports goal-directed behavior, while acetylcholine facilitates neuroplasticity and sharpens attention.

Improved White Matter Integrity: Superior white matter tracts strengthen communication between brain regions, allowing faster, more efficient information processing that aids adaptability and strategic thinking.

Increased Neurogenesis in the hippocampus supports memory formation and emotional resilience. Elevated Brain-Derived Neurotrophic Factor (BDNF) levels further facilitate neuroplasticity and cognitive resilience.

[+3] Hyper-Cognitive Integration

Hyper-cognitive integration is a peak cognitive state characterized by seamless logic, creativity, and emotional intelligence integration. Individuals at this stage exhibit heightened multitasking capabilities, profound insights, and an advanced capacity for intuitive reasoning.

Pathway

Administer an RNA therapeutic and combine with cognitive training during neuronal plasticity state. Adopt a neuroprotective lifestyle. As needed, therapies or interventions that optimize the increasing neural network efficiency and synchrony.

Neurobiology

Synchronized Brainwave Activity: Brain regions associated with logical reasoning, emotional processing, and creativity exhibit gamma wave synchronization (30-100 Hz), especially in the prefrontal cortex, and hippocampus. This synchrony enables the integrated processing of complex thoughts and simultaneous multitasking.

Efficient Neural Network Connectivity

A "small-world" network architecture reduces the distance between neurons across regions, facilitating rapid integration of cognitive functions such as memory recall, abstract thinking, and emotional regulation.

Enhanced Dendritic Arborization

Extensive dendritic branching in cortical and subcortical neurons increases synaptic sites, allowing neurons to process multiple inputs, supporting cognitive flexibility, and advanced memory consolidation.

Balanced Neurochemical Modulation

Optimally balanced excitatory (glutamate) and inhibitory (GABA) neurotransmitter levels sustain cognitive efficiency and reduce fatigue. High serotonin and oxytocin levels contribute to enhanced social cognition and emotional intelligence, crucial for interpersonal processing.

[0] Neurotypical

Baseline Cognition

Neurotypical individuals display baseline cognitive functioning, with well-balanced mental abilities across memory, attention, and executive function. Cognitive resilience enables adaptability to stress and efficient information processing.

Neurotypical brains have balanced excitatory and inhibitory neurotransmitter levels, like glutamate and GABA, facilitating stable cognition and emotional regulation. Synaptic density and connectivity across regions like the prefrontal cortex and hippocampus allow for adequate information storage and recall. Typical levels of serotonin and dopamine support emotional stability and motivation, respectively, while acetylcholine aids in attention and learning.

[-1] Anxiety Spectrum Disorder

Anxiety Spectrum Disorder involves chronic stress and heightened neural reactivity, disrupting baseline function. Neurochemical imbalances lead to persistent worry, cognitive fatigue, and reduced flexibility, impacting focus and decision-making.

MCI can have contributing factors, including aging and genetic predisposition. Neuroinflammation has been found to drive conditions affecting vascular health (e.g., hypertension, diabetes).

Early Hippocampal Atrophy

The hippocampus, crucial for memory formation, shows early signs of atrophy and connectivity loss in individuals with MCI, leading to difficulties in forming and retrieving new memories. Reduced Neurogenesis: Neurogenesis in the hippocampus diminishes, reducing brain's capacity to adapt to new information and challenges. Lower levels BDNF eurotrophic factor, contributes to decline.

Impaired Synaptic Plasticity

Synaptic plasticity—the brain's ability to adapt connections for learning—diminishes, reducing cognitive flexibility and impairing memory recall as neurons struggle to form new associations.

Oxidative Stress and Mitochondrial Dysfunction

Increased oxidative stress and mitochondrial impairment in neurons further contribute to energy deficits and cellular aging, undermining cognitive performance.

[-2] Mild Cognitive Impairment (MCI)

Mild Cognitive Impairment (MCI) involves measurable declines in memory, attention, and executive functioning that go beyond age-related cognitive decline but do not yet meet the criteria for dementia. While some individuals with MCI remain stable, others progress to Alzheimer's or other forms of dementia.

MCI can result from various factors, including aging, genetic predisposition, neuroinflammation, and conditions affecting vascular health (e.g., hypertension, diabetes).

Early Hippocampal Atrophy

The hippocampus, crucial for memory formation, shows early signs of atrophy and connectivity loss in individuals with MCI, leading to difficulties in forming and retrieving new memories.

Reduced Neurogenesis

Neurogenesis in the hippocampus diminishes, reducing the brain's capacity to adapt to new information and challenges. Lower levels of BDNF, a neurotrophic factor, contribute to this decline.

Impaired Synaptic Plasticity

Synaptic plasticity—the brain's ability to strengthen or weaken connections in response to learning—is reduced. This limits cognitive flexibility and impacts memory recall, as neurons become less responsive to forming new associations.

Oxidative Stress and Mitochondrial Dysfunction

Increased oxidative stress and mitochondrial dysfunction lead to energy deficits and accelerated cellular aging, impairing cognitive performance.

[-3] Alzheimer's Disease (AD)

Description

Alzheimer's Disease is a severe neurodegenerative disorder characterized by substantial memory loss, language deficits, and impaired executive functions. Early symptoms include memory lapses and spatial disorientation, which progressively worsen to affect daily functioning and self-care.

Pathway

Alzheimer's is driven by a complex interplay of genetic factors (e.g., APOE-e4 allele), aging, and neuroinflammation, resulting in severe structural and functional brain damage.

Hippocampal Hyperactivity and Atrophy

Early stages of AD show heightened hippocampal activity, which initially compensates for cognitive deficits but eventually leads to neuronal exhaustion and atrophy, impairing memory and spatial navigation.

Widespread Neuroinflammation

Chronic neuroinflammation exacerbates neuronal damage, with microglial cells (brain immune cells) becoming overactive and releasing pro-inflammatory cytokines, accelerating neuronal death.

Disrupted Cortical Networks

Functional connectivity within and across key brain regions, like the hippocampus, prefrontal cortex, and temporal lobes, declines as neurons die. This disruption leads to severe cognitive deficits in memory, reasoning, and executive function.

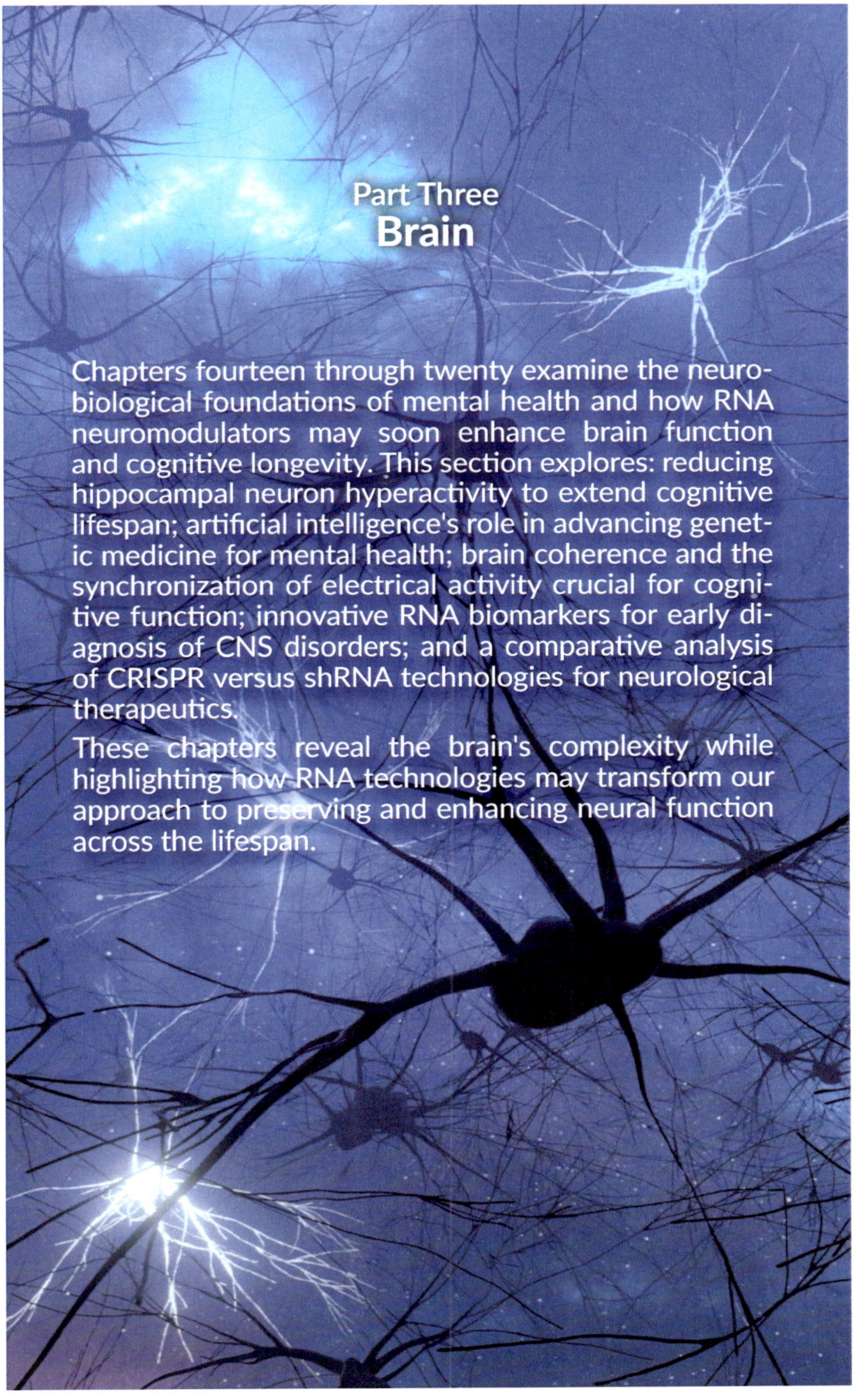

Part Three
Brain

Chapters fourteen through twenty examine the neurobiological foundations of mental health and how RNA neuromodulators may soon enhance brain function and cognitive longevity. This section explores: reducing hippocampal neuron hyperactivity to extend cognitive lifespan; artificial intelligence's role in advancing genetic medicine for mental health; brain coherence and the synchronization of electrical activity crucial for cognitive function; innovative RNA biomarkers for early diagnosis of CNS disorders; and a comparative analysis of CRISPR versus shRNA technologies for neurological therapeutics.

These chapters reveal the brain's complexity while highlighting how RNA technologies may transform our approach to preserving and enhancing neural function across the lifespan.

14. Long-lasting Genetic Therapeutics for Debilitating Neurological Conditions

By Dean Radin

Millions of people suffer from mild cognitive impairment (MCI), a major risk factor for dementia and Alzheimer's disease that is commonly associated with anxiety. Unfortunately, there are no disease-modifying treatments for MCI or dementia, and the selective serotonin-reuptake inhibitors (SSRIs) typically used to treat anxiety have limited effectiveness, inadequate response rates, and adverse side effects.

Targeted Manipulation of Neural Networks

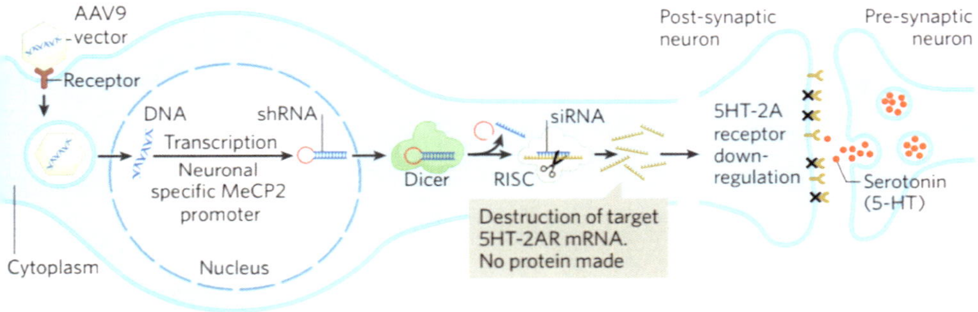

Fg. 1 | Adeno-associated virus 9 (AAV9) vectors enter cells via receptor-mediated uptake, shedding their coats to release DNA plasmids. The DNA, driven by a neuron-specific promoter, produces short-hairpin RNA (shRNA) targeting 5-hydroxytryptamine receptor 2A (5-HT2a) messenger RNA (mRNA). The synthesized shRNA, processed into small-interfering RNA (siRNA), works with the RNA-induced silencing complex (RISC) to degrade 5-HT2a mRNA, reducing receptor synthesis.

The absence of effective treatment options creates a pressing need for novel, more effective, and safer therapeutic interventions for these debilitating disorders. Enter Cognigenics Inc., a biotechnology company specializing in advanced gene editing via intranasal delivery, focusing initially on treatments for MCI with chronic anxiety.

The Platform

A substantial body of research points to hyperactivity in the amygdala and hippocampal regions of the brain in anxiety and memory-associated disorders, respectively. Harnessing genetic-neuroengineering techniques, Cognigenics uses short-hairpin RNA (shRNA) and clustered regularly interspaced short-palindromic repeats (CRISPR)/CRISPR-associated protein 9 (Cas9) genome-editing tools to selectively target and modulate relevant neuronal receptors, enabling precise manipulation of the neural networks associated with anxiety and memory.

The genetic cargo is delivered intranasally to the brain via adeno-associated virus (AAV) vectors. The olfactory highway to the brain effectively bypasses the blood–brain barrier, while the use of AAV vectors ensures safe and efficient transportation of genetic material directly to neurons, facilitating gene-specific silencing in the limbic system.

Cognigenics' initial therapeutics are designed to selectively downregulate the 5-hydroxytryptamine receptor 2A (5-HT2a) post-synaptic neuronal receptors—a serotonin receptor subtype—in the brain via RNA or DNA genetic edits. The company's current focus is on treating MCI and chronic anxiety, to meet the needs of a potentially huge market3. The underlying mechanism uses RNA interference (RNAi) to degrade 5-HT2a messenger RNA (mRNA) and reduce 5-HT2a receptor levels (Fig. 1). The observed behavioral effects are significantly reduced anxiety and enhanced memory.

The company has confirmed these neuronal and behavioral outcomes via immunohistochemistry to trace where the genetic cargo diffuses in the brain; multi-electrode array electrophysiology to assess changes in neuronal performance; polymerase chain reaction (PCR) analyses to verify the genetic edits; and standard behavioral tests in mice and rats. In multiple experiments, the therapeutics resulted in up to 36% reduction in anxiety and 104% improvement in memory (as compared to controls)—surpassing the efficacy of standard first-line pharmaceutical interventions for anxiety and memory disorders, and without discernible side effects. In addition, experiments with appropriately revised genetic edits in human neurons in vitro show similar downregulation effects, indicating that the delivery technique and the expected neuronal regulation effects will be applicable to humans.

"These results suggest that our therapeutics could significantly reduce the burden of MCI and anxiety, paving the way for the potential reduction or eradication of these conditions at their neuronal source," said co-founder and chairman Dean Radin, who anticipates that clinical trials may begin early in 2026.

SSRIs necessitate daily dosing, while most contemporary gene-based interventions for neurodegenerative conditions require either direct injection into the cerebrospinal fluid or intravenous administration, risking nerve damage and potential side effects from systemic distribution, explained Troy Rohn, director of preclinical studies. "In contrast, our approach uses intranasal delivery, which we now know can precisely target specific neuronal receptors and yield lasting therapeutic effects—an innovation in neuropsychiatric treatment that is also likely to improve patient compliance," said Rohn. "With millions of people potentially benefitting, the projected cost per person is estimated to be comparable to existing SSRI drugs."

Moreover, the versatility of this ground-breaking delivery platform is expected to enable customized therapeutic options aimed at addressing an extensive array of neurological conditions or behaviors, including chronic depression, dementia, and Alzheimer's disease.

Partnering Aspirations

Unlike most pharmaceutical companies, which are typically characterized by massive overhead costs and slow, risk-averse processes, *Cognigenics* operates with minimal overheads and is known for its rapid, entrepreneurial approach. The company achieves its efficiency by working closely with contract research and manufacturing organizations. This not only significantly slashes research and development (R&D) time and expenses, but also provides the company with a way to pursue multiple projects in parallel.

"We are addressing the complex challenges of MCI, anxiety, and, potentially, a broad spectrum of central nervous system (CNS) disorders at their root. Our intranasal genetic-medicine delivery platform in development, provides a new class of therapeutics that is both patient-friendly and highly effective," said Radin.

"We stand at the forefront of a new era in mental-health treatments, offering hope and tangible solutions to those in need."

15. Cognitive Longevity & Hippocampal Neuron Hyperactivity

The intricate relationship between neural activity and human lifespan has become a focal point in neuroscience and longevity research. In particular, dysregulation of hippocampal activity has been implicated in age-related cognitive decline, with evidence linking hippocampal hyperactivity to synaptic dysfunction, impaired memory consolidation, and increased vulnerability to neurodegenerative processes.

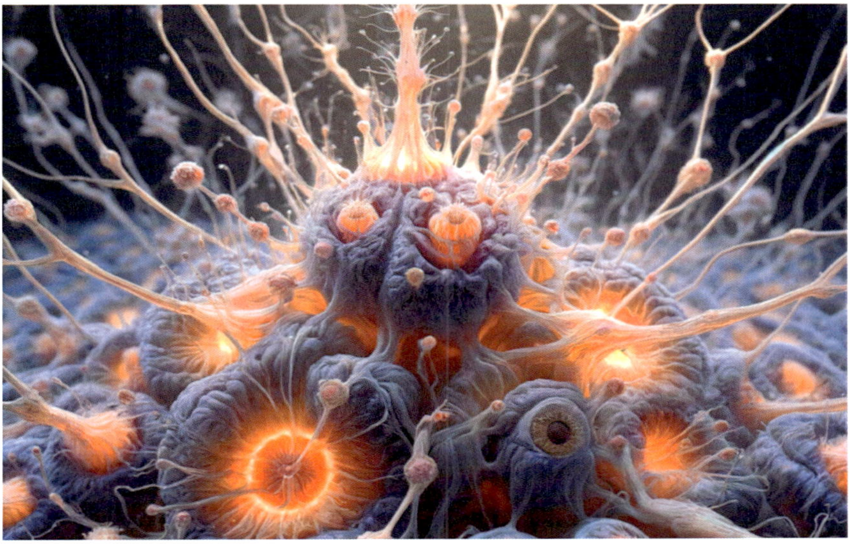

Hyperactive receptors in the limbic brain region. Rendering.

Understanding the mechanistic underpinnings of this hyperactivity is essential for developing targeted interventions that promote neuroprotection and cognitive resilience.

Emerging RNA-based therapeutics designed to modulate aberrant hippocampal activity are now within sight for precision neuroengineering approaches. These interventions leverage short hairpin RNA (shRNA) to restore neural homeostasis, enhance synaptic plasticity, and mitigate the progression of neurodegenerative disorders by selectively silencing maladaptive gene expression. By targeting molecular drivers of hippocampal dysfunction, these therapies may preserve cognitive integrity and extend the health span in aging populations.

REFERENCES I

Aitta-aho T, et al. Attenuation of novelty-induced Gria1-/- mice hyperactivity by cannabidiol and hippocampal inhibitory chemogenetics. *Front Pharmacol.* 2019;10:309. doi:10.3389/fphar.2019.00309

Explores role of hippocampal inhibitory mechanisms in modulating hyperactivity and potential therapeutic strategies for cognitive dysfunction.

Bakker AB, Krauss GL, et al. Reduction of hippocampal hyperactivity improves cognition in amnestic mild cognitive impairment. *Neuron.* 2012;74(3):467-474. doi:10.1016/j.neuron.2012.03.023

Establishes that reducing hippocampal hyperactivity enhances cognitive function in individuals with amnestic mild cognitive impairment.

Busche MA, Chen X, et al. The critical role of soluble amyloid-β for early hippocampal hyperactivity in a mouse model of Alzheimer's disease. *Proc Natl Acad Sci USA.* 2012;109(22):8740-8745. doi:10.1073/pnas.1206171109

Links soluble amyloid-β accumulation to hippocampal hyperactivity in early Alzheimer's disease (AD), demonstrating its impact on neuronal dysfunction.

Huijbers W, et al. Tau accumulation in clinically normal older adults is associated with hippocampal hyperactivity. J Neurosci. 2018;39(3):548-556. doi:10.1523/jneurosci.1397-18.2018

Shows hippocampal hyperactivity is correlated with early tau accumulation in cognitively normal older adults.

II. Mechanisms and Consequences of Hippocampal Neuron Hyperactivity

Hippocampal hyperactivity, characterized by excessive neuronal firing, has been associated with age-related cognitive impairment, mild cognitive impairment (MCI), and Alzheimer's disease (AD). This excessive excitatory activity disrupts neural circuits responsible for memory encoding and retrieval, leading to cognitive dysfunction.

Key Factors Contributing to Hippocampal Hyperactivity: Hippocampal hyperactivity is closely linked to tau pathology and amyloid accumulation. Elevated cerebrospinal fluid (CSF) tau levels are associated with increased neuronal excitability, leading to memory deficits and accelerated cognitive decline.

Aging-related neurochemical changes also play a significant role.

Dysregulation of glutamate, the brain's primary excitatory neurotransmitter, and impaired inhibitory GABAergic signaling disrupts the delicate balance of neural activity, contributing to excessive hippocampal firing. Excitotoxicity and synaptic dysfunction further exacerbate cognitive impairments. Persistent hyperactivity damages synapses, weakens neural circuits and impairs memory formation and retrieval.

Recent neuroimaging studies reveal that CA3 hippocampal hyperactivity is an early biomarker of cognitive decline. Addressing this imbalance is critical for maintaining cognitive function and promoting brain longevity.

REFERENCES II

Mechanisms and Consequences of Hippocampal Neuron Hyperactivity

Aitta-aho T, et al. Attenuation of novelty-induced Gria1−/− mice hyperactivity by cannabidiol and hippocampal inhibitory chemogenetics. *Front Pharmacol.* 2019;10:309. doi:10.3389/fphar.2019.00309

Explores the role of hippocampal inhibitory mechanisms in modulating hyperactivity and potential therapeutic strategies for cognitive dysfunction.

Bakker AB, Krauss GL, et al. Reduction of hippocampal hyperactivity improves cognition in amnestic mild cognitive impairment. *Neuron.* 2012;74(3):467-474. doi:10.1016/j.neuron.2012.03.023

Establishes that reducing hippocampal hyperactivity enhances cognitive function in individuals with amnestic mild cognitive impairment (aMCI).

Busche MA, Chen X, et al. The critical role of soluble amyloid-β for early hippocampal hyperactivity in a mouse model of Alzheimer's disease. Proc Natl Acad Sci USA. 2012;109(22):8740-8745. doi:10.1073/pnas.1206171109

Links soluble amyloid-β accumulation to hippocampal hyperactivity in early Alzheimer's disease (AD), demonstrating its impact on neuronal dysfunction.

Huijbers W, et al. Tau accumulation in clinically normal older adults is associated with hippocampal hyperactivity. J Neurosci. 2018;39(3):548-556. doi:10.1523/jneurosci.1397-18.2018

Shows that hippocampal hyperactivity is correlated with early tau accumulation in cognitively normal older adults, suggesting a preclinical marker of AD.

III. The Role of Brain Health in Longevity and Aging

As individuals age, the brain undergoes complex structural and functional modifications, including synaptic remodeling, neuroinflammatory changes, and altered neurotransmitter dynamics. They all contribute to cognitive decline. One of the key pathological hallmarks of aging-related cognitive dysfunction is hippocampal hyperactivity, which has been directly implicated in memory impairment, increased susceptibility to neurodegenerative diseases, and progressive neural circuit dysregulation.

Research indicates that selectively downregulating 5-HT2a receptors in specific neuronal populations within the hippocampus may mitigate excessive excitatory activity, thereby reducing hippocampal hyperactivity. This targeted intervention has the potential to preserve synaptic integrity, enhance cognitive resilience, and extend health span by modulating serotonergic signaling in a manner that promotes optimal neural function. Addressing dysregulated excitatory-inhibitory balance in aging circuits may represent a novel strategy for delaying cognitive decline and promoting longevity.

Neurobiological Drivers of Aging and Cognitive Decline

Chronic stress and neuroinflammation play roles in hippocampal hyperactivity. Prolonged stress triggers elevated cortisol levels and disrupts neural regulation, leading to excitatory activity and cognitive impairments (Aitta-aho et al., 2019).

Mitochondrial dysfunction further exacerbates the problem. As neurons age, their ability to generate energy declines, making them more vulnerable to hyperactivity-induced damage and accelerating neurodegeneration (Fang, 2023). Another factor is the loss of synaptic plasticity. Reduced neurogenesis and synaptic loss weaken the brain's ability to adapt and form new memories, impairing cognitive flexibility (Berron et al., 2019).

REFERENCES III

The Role of Brain Health in Longevity and Aging

Fang F. Neuronal hyperactivity in the hippocampus during the early stage of streptozotocin-induced type 1 diabetes in mice. *Neuroendocrinology*. 2023;114(4):356-364. doi:10.1159/000536029

This study highlights the effects of metabolic disorders on hippocampal hyperactivity, emphasizing the broader implications of systemic health on brain longevity.

Leal SL, et al. Hippocampal activation is associated with longitudinal amyloid accumulation and cognitive decline. eLife. 2017;6:e22978. doi:10.7554/*elife*.22978

This research links increased hippocampal activity with progressive amyloid accumulation and cognitive deterioration, reinforcing the connection between brain health and aging.

Talati P, et al. Increased hippocampal CA1 cerebral blood volume in schizophrenia. NeuroImage Clin. 2014;5:359-364. doi:10.1016/j.nicl.2014.07.004

Investigates the relationship between hippocampal hyperactivity and schizophrenia, demonstrating how altered hippocampal function contributes to psychiatric disorders and long-term cognitive health.

Yassa MA, et al. Pattern separation deficits associated with increased hippocampal CA3 and dentate gyrus activity in nondemented older adults. Hippocampus. 2011;21(9):968-979. doi:10.1002/hipo.20808

Connects hippocampal hyperactivity to impaired memory function in aging, suggesting a potential early biomarker of cognitive decline.

IV. The Relationship Between Hippocampal Hyperactivity and Longevity

Experimental research in animal and human models confirms the strong link between hippocampal hyperactivity and accelerated cognitive decline.

> "Unveiling the intricate connections between hyperactive neural patterns in the hippocampus and aging could unlock transformative breakthroughs in safeguarding cognitive vitality and extending the human health span." — John Mee, Founder, Cognigenics.

Scientific Findings

These findings suggest that targeting hippocampal hyperactivity would enhance cognitive longevity and overall quality of life. Older adults with high hippocampal activity exhibit more tau accumulation, which correlates with memory deficits and an increased risk of cognitive decline.

Amyloid-β plaques arise from and contribute to early neuronal hyperactivity, disrupting normal hippocampal function and accelerating the onset of Alzheimer's disease.

Persistent hippocampal overactivity worsens neurodegeneration by damaging synapses and impairing memory consolidation, further accelerating cognitive deterioration.

REFERENCES IV

The Relationship Between Hippocampal Hyperactivity and Longevity

Talati P, et al. Increased hippocampal CA1 cerebral blood volume in schizophrenia. *NeuroImage Clin.* 2014;5:359-364. doi:10.1016/j.nicl.2014.07.004

Investigates the relationship between hippocampal hyperactivity and schizophrenia, demonstrating how altered hippocampal function contributes to psychiatric disorders and long-term cognitive health.

Yassa MA, et al. Pattern separation deficits associated with increased hippocampal CA3 and dentate gyrus activity in nondemented older adults. *Hippocampus.* 2011;21(9):968-979. doi:10.1002/hipo.20808

Connects hippocampal hyperactivity to impaired memory function in aging, suggesting a potential early biomarker of cognitive decline.

Paumier A, et al. Astrocyte-neuron interplay is critical for Alzheimer's disease pathogenesis and is rescued by TRPA1 channel blockade. Cold Spring Harb Lab. Published online 2021. doi:10.1101/2021.03.29.437466

Examines role of astrocyte-neuron interactions in Alzheimer's and suggests RNA-based interventions targeting TRPA1 as a therapeutic strategy.

V. RNA-Based Precision Medicine Interventions and Therapeutic Approaches

Given the established role of hippocampal hyperactivity in cognitive decline, Scientists are now focusing on developing interventions that restore neuronal balance, protect against excitotoxicity, and enhance synaptic plasticity.

1. Lifestyle and Behavioral Modifications

Regular exercise enhances neurogenesis and helps regulate hippocampal activity, reducing the risk of cognitive decline, while dietary interventions rich in antioxidants and essential nutrients combat neuroinflammation, protecting neurons from age-related damage; complementing these approaches, stress reduction techniques such as mindfulness and meditation lower excessive excitatory activity, promoting neural balance and emotional well-being.

2. Pharmacological Interventions

GABAergic modulators restore inhibitory signaling in the brain, helping to counteract hippocampal hyperactivity and maintain cognitive function, while neuroprotective compounds, including antioxidants and anti-inflammatory drugs, mitigate oxidative stress and reduce neuronal damage, improving overall brain health.

3. **RNA-Based Therapies: A Breakthrough in Cognitive Longevity**

Researchers are pioneering RNA-based therapeutics that precisely modulate hippocampal hyperactivity using short hairpin RNA. In preclinical studies, this approach enables targeted gene silencing, reducing overactive neuronal receptors to restore neural balance. Unlike traditional treatments, shRNA therapy offers long-lasting neuroprotection without systemic side effects while enhancing synaptic plasticity, which supports memory formation and cognitive function.)

Noninvasive Intranasal Delivery System

To optimize treatment delivery, Cognigenics has developed a proprietary intranasal RNA platform that allows direct drug administration to the brain while bypassing the blood-brain barrier. This noninvasive approach may improve patient compliance by eliminating the need for injections and ensures more precise targeting, enhancing the therapeutic efficacy for hippocampal dysfunction.

References V.

RNA-Based Precision Medicine Interventions and Therapeutic Approaches

Talati P, et al. Increased hippocampal CA1 cerebral blood volume in schizophrenia. *NeuroImage Clin.* 2014;5:359-364. doi:10.1016/j.nicl.2014.07.004

Investigates the relationship between hippocampal hyperactivity and schizophrenia, demonstrating how altered hippocampal function contributes to psychiatric disorders and long-term cognitive health.

Yassa MA, et al. Pattern separation deficits associated with increased hippocampal CA3 and dentate gyrus activity in nondemented older adults. Hippocampus. 2011;21(9):968-979. doi:10.1002/hipo.20808

Connects hippocampal hyperactivity to impaired memory function in aging, suggesting a potential early biomarker of cognitive decline.

Chen K. Anesthesia-induced hippocampal-cortical hyperactivity and tau hyperphosphorylation impair remote memory retrieval in Alzheimer's disease. Alzheimers Dement. 2023;20(1):494-510. doi:10.1002/alz.13464

Links anesthesia-induced hippocampal hyperactivity with tau pathology and memory impairment, highlighting potential risks for long-term brain health.

VI. Implications for Longevity and Mental Health

Targeting hippocampal hyperactivity could transform the treatment of neurodegenerative diseases, anxiety spectrum disorders, and age-related cognitive decline. Addressing the root causes of neural dysfunction offers new possibilities for long-term brain health.

Potential Benefits of Modulating Hippocampal Activity

Reducing neuroinflammation is crucial for preventing neurodegeneration and preserving cognitive function. Protecting memory circuits helps maintain synaptic integrity, ensuring that aging brains retain their ability to process and store information. Enhancing resilience to stress also mitigates chronic stress-induced neuronal damage, a key contributor to cognitive decline.

Advancing Cognitive Longevity: Scientists lead the way extending cognitive health span and promoting brain longevity in preclinical studies, neuroplasticity research, and longevity science. Through these advancements, researchers aim to redefine the future of mental health and neuroprotection.

VII. The Future of Cognitive Longevity

Key Innovations and Market Impact: Preclinical compounds, have demonstrated an impressive 100% improvement in memory performance during preclinical trials, highlighting its potential as a revolutionary treatment.

Growing Market: The global brain health and longevity market is projected to surpass $250 billion by 2030.

Redefining Mental Health and Longevity: Combining RNA-based precision medicine with cutting-edge neuroplasticity research, Cognigenics is redefining the future of mental health and cognitive longevity, offering transformative solutions for aging and neurodegenerative diseases.

VIII. Conclusion: A New Era of Brain Longevity

Understanding and modulating hippocampal hyperactivity is essential for extending cognitive longevity and maintaining brain health. With RNA-based therapeutics at the forefront, we are witnessing a paradigm shift in how we approach aging, memory loss, and mental well-being.

Researchers are rapidly bridging the gap between genetic medicine and longevity science to unlock new possibilities for a healthier, longer life.

16. Artificial Intelligence and Genetic Medicine

Artificial Intelligence (AI) and genetic medicine have made remarkable advancements in recent years, and their combination is proving revolutionary in the development of RNA neuromodulators. Genetic researchers are expediting research, testing, and translation to improve RNA-based therapies, opening new frontiers for addressing neurological and psychiatric conditions.

```
GTGTTGTAGTCATGAAAGGGTGGGTTTTGATTGACCCATTTGTGTTTTTGCT
GTTGTTGACTGACCTTGCTTTTGATCGGCTGCTTATGCATGTGGGGAA
GTTGACTGTAGTTTGGGAGTGTGGAAGCTGGTGAAGGGATCTGGGAGTAC
AGCTCGATTGCTTGTGCTGGGGTCGACTGTGTTGGGAGAGTGGTTTCTGATGTCT
TCTTGATCTGCTTTGACTGTAGTGTGGAAGTTGACATGGTTTCTGGAAGAGCTG
GAGCTGATGACGTTGACCTGTGGTGCTGGTTGGAGAGGAGGAAGGGAATGTCT
GTGGTGCTGGGTAGTTGACTGTGGAGCTGGAAGTGACTGCTGGAAGTGTTAGGT
GTTGGGAGGATTGGGGGAGGTCTGGTGGTGCTGGGGTTGACTGTAGGAGTGGAA
GAGGCTGGAAGTGGGTTTGGGAGGGTCTGTAGGTGGTGCTGGAAGGAGGGGGTG
TAGGGTGCGAGTTGACTGTGGAGGTGGGGGTGTGGTGGTGGGTTGACTGTGGAA
GGGGAGGGGGTGTTGGTGCTGGGTTGACTGTGGAGGTGGAGGGGGTGTGGTGCT
GGGTTGACTGTGGAAGGGGAGGGTGTTGGTGCTGGGTTGTTACTGTGGAGG
ATGAACTTCATCATCATTGATCTGCTTTGGCATAGTTTGAGAGATTAAGGAATGGCCT
CATTGATCAGTGGCATTGTTATTTTGGCAACCTGGGGTCATGAATGGTTTTTCT
TTGCAAGTGGGTGCAGGAAAGGAAGTTCGGTGTTGAGGCTTTCTGGAACATTTT
GACATGCGGCTATGGACTCCGTGTTTGACTTTTGGGTTTGGACTTCGTGGTGAA
GGGAGGGGGTGTGGTGCTGGGTTGACTGTGGAGGGGGAGGGGGTGTTGGTGCT
GGGTTGACTGTGGAGGTGGAGGGGGTGTGGTGCTGGGTTGACTGTGGAAGGGG
GAGGGGGTGTTGGTGCTGGGTTGACTGTGGATGGGGGATGGGGGTGTTGGTGCTG
GGTTGACTGTGGAGGTGGAGGGGGTGTGGTGCTGGGTTGACTGTGGAAGGGGAT
GAACTTCATCATCATTGATCTGCTTTGGCATAGTTTGAGAGATTAAGGAATGGCCT
CATTGATCAGTGGCATTGTTATTTTGGCAACCTGGGGTCATGAATGGTTTTTCT
TTGCAAGTGGGTGCAGCGAAAGGAAGTTCGGTGTTGAGGCTTTCTGGAACATTTT
GACATGCGGCTATGGACTTCGTGTTTGACTTTTGGGTTTGGACTTCGTGGTTGAT
GAGGGGGTGTTGGTGCTGGGTTGACTGTGGAGGGGGAGGGGGTGTTGGTGCTG
GTTGACTGTAGTTTGGGAGTGTGGAAGCTGGTGAAGGGATCTGGGAGTACGGG
GTGTTGGTGCTGGGTTGCTGTGGGGGAGGGTGTTGGGGGTTGTGCTGGGTTGACT
```

Short excerpt from the start of the coding region of the HTR2A gene

The Dawn of Genetic Medicine

The DNA structure, first defined in the 1950s, is now actualizing visions of advanced genetic medicine. Many of the complexities of heredity and hereditary illnesses are being discovered as a result. In 2003, scientists completed the Human Genome Project and created the first map of the human genetic code, essentially generating a blueprint of human DNA. These discoveries improved our understanding of genetic disorders and the ability to tailor treatment to particular patients individually.In the following years and particularly in the last decade, RNA treatments have emerged as a promising subject, with applications including mRNA vaccines,

RNA interference (RNAi), and antisense oligonucleotides (ASOs). These new therapeutic tools use RNA's essential function in translating genetic information into proteins, allowing for molecular interventions at the root of disease processes. Despite its potential, however, RNA-based therapeutics faced considerable obstacles, such as RNA molecule stability, effective distribution to target cells, and reducing off-target effects.

AI and ML as Transformative Technology

Artificial intelligence, or machine learning (ML), are systems that learn from and forecast data. AI applications in health care have evolved from simple data management to complex predictive analytics and personalized medicine. AI models have been used to analyze vast datasets, predict molecular structures, and simulate biological processes, making them valuable tools in genetic research.

The use of AI in genetic research has transformed the analysis of data and the practice of gene editing. Artificial intelligence is capable of effectively managing and analyzing the large amounts of data produced by high-throughput sequencing technologies. Deep learning algorithms have the potential to perceive patterns in DNA sequences beyond the understanding of humans. For example, AI enhances CRISPR-Cas9 technology for gene editing by forecasting the most advantageous guide for RNA sequences, diminishing unintended effects on non-target genes, and augmenting the precision of the editing process.

Expediting Progress with Artificial Intelligence

Artificial intelligence serves several roles in RNA drug development. Its systems can identify RNA targets by analyzing genetic data and predicting the biological consequences of altering specific RNA sequences. In this way, the precision of RNA structure predictions is crucial in advancing effective medicines. CONTRAfold and other machine learning techniques enhance the accuracy of these forecasts. Artificial intelligence also enables evaluating the effective distribution of RNA molecules to target cells by improving delivery methods, such as lipid nanoparticles.

Case Studies and Success Stories

Moderna's rapid development of mRNA vaccines for COVID-19 illustrates the importance of artificial intelligence. AI-powered models aided in identifying viral RNA sequences, predicting their structural behavior, and optimizing mRNA constructions. Moderna's integration has boosted vaccine research, leading with unprecedented speed and efficacy in combating the pandemic.

Alnylam Pharmaceuticals uses artificial intelligence to produce RNAi-based therapeutics. AI algorithms can identify RNA interference (RNAi) targets, calculate silencing efficacy, and enhance RNAi medications. This technique illustrates artificial intelligence's potential to revolutionize RNA therapeutics.

Ethical and Practical Considerations

Artificial intelligence will become an embedded element of most facets of genetic research. In advance of that, it would be sensible to address potential ethical conflicts. Ensuring the privacy and protection of data, particularly genetic information, is of the utmost importance. Health care AI models need to maintain objectivity to prevent inconsistencies in treatment results. For artificial intelligence systems to effectively reflect population genetic variation, they need to undergo training using a diverse variety of inputs, which has not always been the case and thankfully is now becoming a mainstream concern.

Future Directions and Potential Solutions

The prospects for AI in genetic medicine are promising, in part because of advancements such as quantum computing and advanced neural networks. These methods are eventually becoming commercially available and are poised to extend AI capabilities. Partnerships between AI experts and genetic researchers are productive for uncovering novel capabilities.

As an example, the success of AlphaFold in forecasting protein structures showcases the capacity of interdisciplinary research. AlphaFold is a groundbreaking artificial intelligence system developed by DeepMind, a subsidiary of Alphabet Inc. It's designed to predict the three-dimensional structure of proteins based on their amino acid sequences, effectively addressing the long-standing "protein folding problem" in structural biology. This is crucial because a protein's structure largely determines its function in biological processes.

The system employs deep learning techniques, particularly attention-based neural networks, trained on a database of known protein structures and amino acid sequences. AlphaFold's impact extends to various areas of scientific research and drug development. Its predictions can potentially accelerate research in drug discovery, disease mechanism understanding, and the development of new industrial enzymes. In 2021, DeepMind released the source code for AlphaFold and created a database of predicted structures for nearly all known proteins, making this information freely available to the scientific community.

While highly accurate for many proteins, AlphaFold still faces challenges with certain complex proteins, particularly those that are part of larger complexes or undergo significant conformational changes. Despite these limitations, AlphaFold represents a significant advancement in computational biology and demonstrates the potential of AI to solve complex scientific problems, earning recognition as one of the most significant scientific breakthroughs in recent years.

AI can accelerate the creation of genetic medicine and optimize drug formulation and delivery. Integrating AI tools opens a new therapy era, offering optimism for what previously were considered untreatable conditions and, as a result, may revolutionize health care and enhance patient outcomes.

RNA Neuromodulators: Advancing Treatment Approaches

In this complex interplay of genetic medicine with AI, RNA neuromodulators show growing potential for treating several unmet needs in human health, like neuroinflammation. Ongoing studies focus on present difficulties and investigate novel remedies. Nanoparticles and exosomes show potential as effective means for delivering RNA treatments across the blood-brain barrier (BBB) and for selectively targeting particular areas of the brain, improving the effectiveness and accuracy of drug delivery.

Viral vectors, such as adeno-associated viruses (AAVs), can be engineered to transport RNA molecules directly to neuronal cells, circumventing the BBB and guaranteeing precise brain target delivery of therapeutic treatment. Currently, it is possible to administer mRNA vaccines using lipid nanoparticles (LNPs). LNPs are artificially engineered microspheres whose design is grounded on our current knowledge of naturally occurring similar biological objects called "exosomes," which may also carry RNA drugs to the brain. Simultaneously, they enhance cellular uptake and protect RNA molecules from destruction. Exosomes are endogenous vesicles that may be modified to transport RNA treatments and target certain cell types. Exosomes provide benefits due to their biocompatibility and ability to traverse the BBB.

RNA neuromodulators can be greatly improved in terms of stability and effectiveness via the use of chemical modifications and delivery mechanisms. Modifications such as 2'-O-methyl and phosphorothioate backbones provide protection to RNA molecules against nucleases and decrease immunological activation. The addition of 2'-O-methylation to RNA molecules enhances their stability and lowers immunological recognition, hence increasing their therapeutic potential. Phosphorothioate backbones also enhance the stability of RNA molecules, rendering them more resilient to enzyme destruction. Lipid nanoparticles, formed by enclosing RNA within lipid nanoparticles, shield RNA from degradation and enable its transportation to specific cells, thus augmenting the effectiveness of therapy.

Precision Medicine and Personalized Approaches

RNA neuromodulators can be customized to an individual's genetic profile. For select RNA medications, this has the potential to create personalized versions of the treatments. These can improve the effectiveness of the therapy while reducing adverse reactions. Artificial intelligence and machine learning are gaining significance in developing tailored medications and identifying pertinent RNA targets.

RNA neuromodulators show great potential to treat diseases characterized by inflammation of the limbic system. These treatments focus on distinct inflammatory pathways and may provide accurate treatments with few adverse effects. To fully harness their capabilities, it is necessary to tackle the issues of delivery and stability.

Effective integration of RNA neuromodulators into clinical practice for patients with neuroinflammatory disorders will soon be possible. The availability of these genetic therapeutics is being made possible by ongoing research, multidisciplinary collaboration, and technical improvements. These initiatives provide fresh hope to those suffering from neuropsychiatric and neurocognitive conditions, representing one of the most promising frontiers in modern medicine's quest to address some of humanity's most challenging health conditions.

17. An Inquiry Into Brain Coherence

Brain coherence is the synchronization of electrical activity between different brain parts, measured through techniques like electroencephalography (EEG). This coherence reflects the harmonious communication and connectivity between brain regions, essential for integrating information and executing complex cognitive and motor functions efficiently. The effects and significance of brain coherence span various aspects of cognitive functioning and mental health.

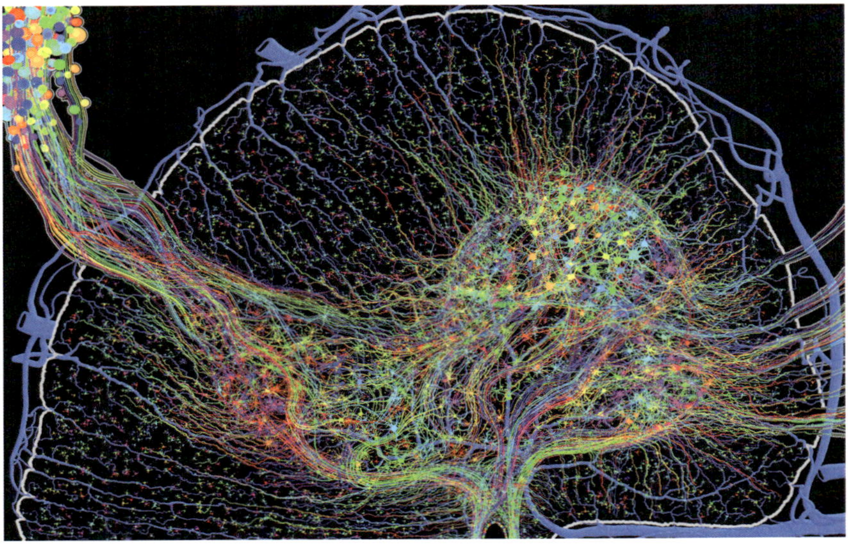

Mapping Gray Matter, Greg Dunn

Cognitive Performance

Increased coherence, especially in particular frequency bands, is often associated with improved cognitive functions, including memory, attention, and problem-solving skills. Coherent brain activity allows for more effective neural communication, which is crucial for quickly and accurately processing information.

Brain coherence plays a vital role in learning and memory. Enhanced synchronization between the hippocampus (critical for memory formation and retrieval) and other brain regions can facilitate consolidation of new memories and the integration of new learning with existing knowledge.

Emotional Regulation

Coherence between emotional processing centers (like the amygdala) and cognitive control centers (like the prefrontal cortex) is vital for emotional regulation. Effective connectivity helps in managing emotional responses and is critical for mental health.

Consciousness and Awareness

Some theories suggest that a certain level of global brain coherence is necessary for consciousness and self-awareness. This global coherence might integrate sensory inputs and mental processes into a unified conscious experience. Practices like meditation and mindfulness have been shown to increase brain coherence. This heightened coherence is associated with deep relaxation, reduced anxiety, and improved stress management. Long-term meditation practitioners often show significantly greater brain coherence than non-practitioners.

Neurodevelopmental Disorders and Neurodegenerative Conditions

Abnormalities in brain coherence have been observed in various neurodevelopmental disorders (such as autism spectrum disorders) and neurodegenerative diseases (such as Alzheimer's disease). These abnormalities can manifest as either hypercoherence (excessively synchronized activity that may hinder the brain's flexibility) or hypocoherence (reduced synchronization that can impair cognitive functions).

Mental Health Disorders

Changes in brain coherence patterns have been linked to several mental health disorders, including schizophrenia, bipolar disorder, major depressive disorder, neurodevelopmental disroders, and neurodegenerative disorders. These changes may reflect disruptions in the standard patterns of neural communication and brain network connectivity, contributing to the symptoms of these conditions.

Brain Coherence

Brain coherence is a fundamental aspect of neural function that impacts a wide range of cognitive abilities, emotional processes, and overall mental health. Understanding and modulating brain coherence can offer potential pathways for enhancing cognitive performance and mental well-being, and treating various neurological and psychiatric disorders.

How Is Brain Coherence Disrupted?

Brain coherence, the synchronization of electrical activity across different brain regions, can be disrupted by various factors, ranging from neurodevelopmental disorders to environmental influences. Disruptions in brain coherence can lead to difficulties in cognitive processes, emotional regulation, and overall brain function.

Causes and Mechanisms of Brain Coherence Disruption

Neurodevelopmental Disorders: Conditions like autism spectrum disorder (ASD) and attention deficit hyperactivity disorder (ADHD) can feature altered patterns of brain coherence. These disruptions can contribute to cognitive and behavioral characteristics, such as difficulties with attention, communication, and social interaction. Neurodegenerative diseases like Alzheimer's and Parkinson's disease disrupt brain coherence. As they progress, neural communication within and between brain regions deteriorates, affecting memory and other cognitive functions.

Mental Health Conditions: Schizophrenia is often associated with both increased and decreased coherence in different brain regions, reflecting the complex nature of the disorder. Such alterations might contribute to symptoms like hallucinations, delusions, and disorganized thinking. Depression and anxiety change the brain's coherence, particularly involving the frontal regions of the brain. These changes relate to difficulties in emotional regulation and processing emotional information.

Brain Injury: A traumatic brain injury from impact to the head can cause immediate and long-term changes in brain coherence, affecting cognitive functions, emotional regulation, and physical coordination. The severity of coherence disruption often correlates with the extent of the injury. Strokes can disrupt brain coherence by damaging brain tissue, thus impairing the affected regions' ability to communicate effectively with the rest of the brain. The impact on coherence often depends on the stroke's location and severity.

Substance Use: Alcohol and specific drugs can lead to changes in brain coherence. These substances alter neural connectivity and communication, and excessive use can lead to cognitive impairments and profound emotional disturbances.

Environmental and Lifestyle Factors: Stress over a long period can affect brain coherence, particularly reducing connectivity in areas involved in emotional regulation. This may contribute to heightened anxiety, depression, and decreased cognitive function as a result.

Sleep deprivation can disrupt normal patterns of brain coherence, affecting attention, memory, and decision-making processes.

Aging: Normal aging is associated with changes in brain coherence, reflecting alterations in neural connectivity and communication. While some changes may be part of the healthy aging process, they can also increase vulnerability to cognitive decline and dementia.

Understanding how brain coherence is disrupted can offer insights into the underlying mechanisms of various conditions and lead to better-targeted interventions. Techniques such as neurofeedback, cognitive behavioral therapies, and certain medications aim to normalize or improve coherence patterns, potentially alleviating some symptoms of disrupted brain coherence.

Modulation of Neural Networks

The 5-HT2a receptor, a subtype of the serotonin receptor, plays a crucial role in various neural processes related to mood, cognition, and perception. It modulates neurotransmission across various neural circuits, influencing brain coherence and synchronizing electrical activity between brain parts. Here's what we know about the involvement of the 5-HT2a receptor in brain coherence: the 5-HT2a receptor is widely distributed in the brain, particularly in regions critical for higher cognitive functions, such as the prefrontal cortex. Activation of these receptors can modulate the activity of neural networks, potentially influencing the patterns of brain coherence. By affecting the excitability and connectivity of neurons, 5-HT2a receptors can alter the synchronization of neural oscillations, critical for coherent brain activity.

Impact on Cognitive Functions: Cognitive processes such as attention, memory, and executive function are underpinned by coherent activity in specific neural networks. The 5-HT2a receptor plays a role in modulating these networks, influencing cognitive performance. For instance, proper functioning of the 5-HT2a receptor system is associated with optimal cognitive flexibility and working memory, which depend on the coherent activity of neurons in the prefrontal cortex.

Role in Psychiatric Disorders: Alterations in the function or expression of 5-HT2a receptors have been implicated in various psychiatric conditions, such as schizophrenia, depression, and anxiety disorders, which are characterized by disruptions in brain coherence. In schizophrenia, for example, abnormal 5-HT2a receptor activity is associated with disrupted connectivity and coherence in brain networks, contributing to symptoms such as hallucinations and thought disorder. Therapeutic agents targeting 5-HT2a receptors can help restore standard patterns of brain coherence and improve psychiatric symptoms.

Psychedelics and Brain Coherence: Psychedelic compounds, such as psilocybin and LSD, exert their profound effects on consciousness and perception primarily through agonist activity at 5-HT2a receptors. Research has shown that these compounds can significantly alter brain coherence patterns, leading to changes in perception, emotion, and cognition. This alteration in brain coherence under the influence of psychedelics has been a subject of interest for understanding consciousness and for therapeutic applications in mental health.

Therapeutic Potential: Modulating 5-HT2a receptor activity holds therapeutic potential for conditions associated with disrupted brain coherence. For instance, certain antipsychotic and antidepressant drugs work by targeting 5-HT2a receptors, among others, aiming to normalize disrupted neural communication and coherence. Additionally, research into psychedelics as therapeutic agents is exploring how induced changes in brain coherence can contribute to lasting improvements in mental health conditions.

The 5-HT2a receptor plays a significant role in brain coherence by influencing neural network modulation, cognitive functions, and psychiatric health. Understanding its involvement offers insights into the neural basis of cognition and behavior and opens avenues for therapeutic interventions working with neural coherence.

Hypothesis

New RNA therapeutics in preclinical development are being developed from the perspective that the therapeutic effect on brain coherence is mediated through its targeted modulation of 5-HT2a receptors, in order to enhance neural synchronization and connectivity in cognitive and emotional regulation networks.

Method of Investigation: To test this hypothesis, a combination of neuroimaging techniques (such as fMRI and EEG) and cognitive/behavioral assessments can be employed in both preclinical models and clinical trials. Specifically, investigating changes in brain coherence patterns, neural connectivity, and mental and emotional performance before and after administration of the RNA therapy would provide insights into its mechanism of action and therapeutic potential.

Expected Outcomes

If validated in the lab, we would expect to observe:

1. Increased synchronization in brain networks associated with cognitive processing/emotional regulation following treatment.
2. Improved performance in cognitive and emotional tasks correlating with changes in brain coherence.
3. Evidence of neuroplastic changes supporting enhanced connectivity and network efficiency.
4. The mechanism of action and its effects on brain coherence, may lead to novel therapeutic strategies targeting the underlying neural dysfunctions.

Benefits of Brainwave Coherence

Brainwave coherence refers to the synchronization of electrical patterns within the brain. When brain waves are coherent, various parts of the brain work together harmoniously and efficiently. This concept is increasingly studied within neuroscience and psychology for its potential benefits on cognitive functions, mental health, and overall well-being. Here are some of the key advantages associated with brainwave coherence:

Improved Cognitive Performance: High levels of brainwave coherence are often linked to better cognitive function. This includes enhanced memory, increased focus and concentration, and faster processing speeds. Coherence has been shown to facilitate more efficient neural communication, improving problem-solving abilities and creativity.

Stress Reduction: Coherence in brainwave patterns can lead to a reduction in stress and anxiety levels, regulating the body's response to stress, promoting a more relaxed and calm state. This can be particularly beneficial in today's fast-paced world, where chronic stress is a common concern.

Emotional Stability: Individuals with higher brainwave coherence tend to exhibit greater emotional resilience and stability. Coherent brainwave patterns can enhance the brain's ability to regulate emotions, making it easier to handle emotional challenges and reducing mood swings.

Enhanced Learning Abilities: Brainwave coherence can improve learning efficiency and retention. It supports the brain's ability to assimilate and integrate new information effectively, making the learning process smoother and more productive.

Better Sleep Quality: Coherence in brainwaves, especially in the alpha and theta ranges, is associated with improved sleep patterns. It can help transition to sleep and increase the quality of REM cycles, leading to better rest and recovery.

Increased Mental Clarity and Awareness: With higher levels of coherence, individuals often report experiencing greater mental clarity and awareness. This heightened alertness can improve decision-making and enhance one's ability to be present and mindful.

Improved Social Skills: Brainwave coherence can also positively impact social interactions. It is associated with improved empathy and communication skills, as it enhances the brain's ability to process and respond to social cues.

Health Benefits: There are indications that brainwave coherence may have various health benefits, including lowering blood pressure and boosting the immune system. These benefits are likely to arise from the stress-reducing effects of brainwave coherence.

Measurable Cerebral Coherence: As detailed above, brain coherence refers to the synchronization of electrical activity between different regions of the brain, a phenomenon that can be measured using techniques like electroencephalography (EEG). This synchronization is crucial for efficient neural communication.

Dimensions of Brain Coherence

Cognitive Performance: Higher coherence in specific frequency bands is linked to improved cognitive functions such as memory, attention, and problem-solving. This suggests that coherent brain activity facilitates efficient neural communication, enhancing the brain's ability to process and integrate information.

Learning and Memory: Coherence between the hippocampus and other brain areas plays a significant role in the consolidation and retrieval of memories. Enhanced synchronization can improve the integration of new information with existing knowledge, crucial for effective learning.

Emotional Regulation: The coordination between emotional centers (e.g., the amygdala) and cognitive control centers (e.g., the prefrontal cortex) is vital for managing emotions. This connectivity aids in the regulation of emotional responses, which is essential for maintaining mental health.

Consciousness and Awareness: Certain theories propose that a baseline level of global brain coherence is necessary for consciousness and self-awareness, integrating sensory inputs and cognitive processes into a unified experience.

Meditation and Relaxation: Practices like meditation are associated with increased brain coherence, leading to states of deep relaxation and reduced anxiety. Long-term practitioners often exhibit greater coherence, suggesting potential long-term benefits.

Mental Health Disorders: Changes in coherence patterns are linked to disorders like schizophrenia, and major depressive disorder, often reflecting disruptions in normal neural communication and connectivity.

Future Research Directions

Understanding and Enhancing Brain Coherence: Due to the significance of brain coherence in different cognitive and emotional functions, more investigation is needed to examine how coherence may be adjusted to enhance mental well-being and cognitive abilities.

Causes of Coherence Disruption: Examine the root causes of altered brain coherence in various neurological and psychiatric disorders. This involves investigating the function of certain neurotransmitter systems, such as the 5-HT2a receptor, and their impact on coherence.

Interventional Strategies: Create and improve strategies that target the restoration or improvement of brain coherence. Possible treatments may encompass neurofeedback, cognitive-behavioral therapy, pharmaceutical medications, as well as innovative genetic or RNA-based therapies.

Longitudinal Studies: Perform longitudinal investigations to understand the progression of brain coherence in different circumstances, especially in relation to neurodegenerative disorders. This will aid in the identification of early indicators of cognitive deterioration and potential opportunities for intervention.

Personalized Medicine: Investigate customized methods for regulating brain coherence, including variations in brain anatomy, activity, and reaction to treatments.

Technological Progress: Employ sophisticated neuroimaging and electrophysiological methods to get a more comprehensive understanding of the dynamics of brain coherence. This encompasses the utilization of high-density EEG and functional MRI techniques in order to more precisely delineate coherence patterns.

Study of Psychedelics: Conduct a more in-depth examination of the impact of psychedelics on brain coherence, specifically focusing on how these chemicals might modify neuronal synchronization and connection. Gaining insight into these impacts might provide valuable guidance for the advancement of novel therapeutic strategies for disorders such as depression and PTSD.

Interfacing with Other Modalities: Investigate the relationship between brain coherence and other physiological and psychological aspects, including hormone levels, sleep patterns, and lifestyle choices. Implementing this all-encompassing strategy may result in the development of more thorough treatment techniques.

Novel Avenues for Mental Well-Being

Gaining a deeper comprehension of brain coherence and how it may be adjusted has great promise for strengthening cognitive abilities, emotional health, and the management of different neurological and psychiatric conditions. By doing focused study and employing inventive therapy methods, we can discover novel avenues for enhancing mental well-being and cognitive abilities.

18. Innovative Longevity Biomarkers for CNS Disorders

Central nervous system (CNS) disorders, such as Alzheimer's, Parkinson's disease, and multiple sclerosis, are difficult to diagnose and treat. Traditional diagnostic techniques frequently fail to detect the early stages of many illnesses, resulting in delayed treatment. RNA-based biomarkers are developing as a game-changing technology, providing precise, noninvasive tools for early detection and disease progression monitoring.

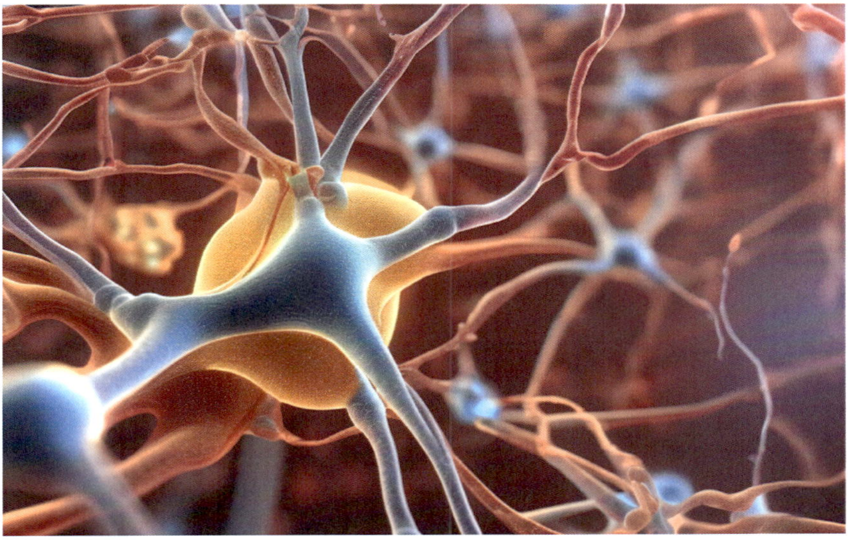

Innovative Bio-markers. Rendering.

The Role of RNA Molecules in CNS Disorders

RNA molecules include several distinct types: messenger RNA (mRNA), microRNA (miRNA), and long non-coding RNA (lncRNA). Each kind serves a distinct function in the regulation of genes and the facilitation of biological activities.

Messenger RNA (mRNA) is the key molecule translating genetic information from DNA to the ribosome, where it is employed for protein synthesis. MicroRNAs (miRNAs) are short non-coding RNA molecules that regulate gene expression by selectively binding to complementary sequences on target mRNAs. This interaction might lead to either the degradation of the mRNA or the inhibition of its translation into proteins. Long non-coding RNAs (lncRNAs) are extended RNA molecules that influence gene expression through multiple mechanisms, including transcription and post-transcriptional processing.

Advantages of RNA Biomarkers

The various coding and non-coding regulatory RNAs provide several benefits over traditional protein biomarkers. Their capacity to identify minor abnormalities in molecules associated with central nervous system disorders is exceptional in precision and sensitivity. MicroRNAs, as non-coding regulatory RNA molecules, maintain structural integrity in various physiological fluids—a characteristic that makes them attractive candidates for noninvasive diagnostics.

Microarrays and next-generation sequencing can identify and investigate both coding and regulatory RNAs with remarkable precision. RNA-based sequencing is emerging as a powerful technique that provides precise and noninvasive tools for early disease detection and enables the development of targeted therapeutic approaches.

Recent research has shown that RNAs may be particularly valuable in the diagnosis and monitoring of central nervous system diseases. For example, studies have revealed specific miRNA signatures associated with the severity and progression of Alzheimer's disease, emphasizing their utility as both diagnostic and prognostic biomarkers. Furthermore, clinical investigations evaluating RNA biomarkers in Parkinson's disease have yielded promising results, with specific microRNAs linked to disease progression and treatment outcomes.

Non-Coding RNAs for Monitoring

MiR-146a is primarily involved in regulating the intrinsic immune response. It acts as a negative feedback regulator of inflammation and targets components of inflammatory signaling pathways. In Alzheimer's disease, elevated levels of miR-146a are present in the hippocampus, a region associated with memory and cognition. This increase may represent a response to chronic inflammation.

MiR-155 is another microRNA that plays a role in modulating the immune response. Both miR-146a and miR-155 target genes that regulate inflammation and immune cell processes. In Alzheimer's disease, these microRNAs are thought to influence neuroinflammation and activate the brain's immune cells to produce pro-inflammatory cytokines. They are associated with the progression of neurodegenerative processes due to their role in sustaining and amplifying inflammatory responses.

The expression levels of microRNAs like miR-146a and miR-155 are closely linked to inflammatory processes in Alzheimer's disease. Monitoring these levels may provide enhanced visibility into disease progression and could be instrumental in assessing the effectiveness of anti-inflammatory therapies.

Exploration of RNA Technologies for Biomarker Discovery

Advanced RNA sequencing technologies, such as next-generation sequencing and single-cell RNA sequencing, are discovering new RNA molecules with biomarker potential. These technologies help identify unique RNA expression patterns associated with central nervous system illnesses. Single-cell RNA sequencing, in particular, provides vital information on cellular diversity and the specific activities involved in disease development.

Data Analysis and Bioinformatics

Integrating data with innovative bioinformatics approaches enhances the process of detecting new RNA biomarkers and confirming existing ones. Machine learning and network analysis have become routine tools for recognizing RNA networks and functional links between them. This enhanced visibility allows for more targeted therapeutic approaches and a better understanding of the molecular mechanisms underlying central nervous system disorders.

Ongoing Clinical Trials

Numerous clinical trials are evaluating RNA biomarkers' effectiveness for central nervous system illnesses. These studies gather empirical evidence supporting RNA biomarkers' diagnostic and prognostic significance while assessing their potential to guide treatment decision-making. Current clinical research is examining the use of microRNAs to monitor the progression of Alzheimer's and Parkinson's diseases and evaluate therapy efficacy.

Before widespread implementation, technical and practical obstacles must be resolved. Consistent techniques will need to be developed with academic consensus. RNA biomarker testing requires consistency and accuracy across different clinical settings. Regulators will likely mandate standardized protocols for collecting, storing, and processing samples to preserve RNA quality.

Ethical and Regulatory Considerations

Genetic data requires special protections. The use of RNA biomarkers raises concerns about patient confidentiality and data safeguarding. To address these issues, regulators can revise current frameworks to ensure the ethical utilization of RNA biomarker data and develop comprehensive clinical implementation guidelines. These guidelines will need to address validation, standardization, and integration of RNA biomarkers into clinical practice.

Future Prospects

Ongoing advancements in RNA sequencing and bioinformatics will increasingly facilitate the discovery of novel RNA molecules—both coding and non-coding—and their potential medicinal applications. Integrating RNA data using multi-omics approaches can improve our understanding of central nervous system disorders and optimize the effectiveness of RNA neuromodulators.

A multi-omics approach comprehensively studies all "omics" data (e.g., proteomics, epigenomics, regulomics) to understand biological systems holistically. Each "omic" represents a distinct layer of biological information, from proteomics (investigating the complete set of proteins produced by a genome, cell, tissue, or organism) to epigenomics (examining chemical changes to DNA and histones that regulate gene function without altering the DNA sequence). By integrating these diverse datasets, scientists can discern interactions and influences across multiple biological layers.

RNA biomarkers hold significant promise for improving the diagnosis and treatment of central nervous system disorders. RNA technologies enable the creation of personalized and effective treatment strategies by providing early, accurate, and noninvasive diagnostic tools. Continued research and innovation in this field have the potential to significantly enhance patient care and outcomes, bringing new hope to individuals affected by central nervous system disorders.

The convergence of RNA biomarker discovery and RNA neuromodulator development creates a powerful synergy—biomarkers enable earlier detection and more precise monitoring, while neuromodulators provide targeted intervention at the molecular level. Together, these advances represent a new frontier in addressing the complex challenges of CNS disorders.

19. Overcoming the Blood-Brain Barrier: A Gateway to RNA Neuromodulation

The blood-brain barrier (BBB) is a protective shield that keeps the brain from harmful substances while allowing vital nutrients to enter safely through its semipermeable structure. It is so effective that it is estimated that approximately 98% of small-molecule drugs and almost 100% of large molecules (like proteins) are prevented from crossing the BBB into the brain. Although this allows the brain to remain pristine and protected, the BBB also poses a significant barrier for the development of CNS therapeutics, including RNA neuromodulators.

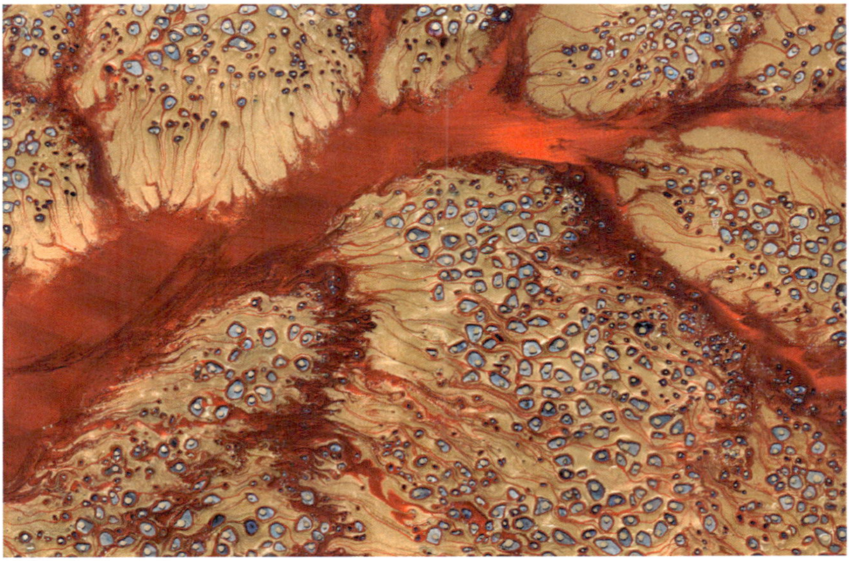

The Blood-Brain Barrier, Greg Dunn

To circumvent the BBB, scientists have turned to an alternative route to the brain using intranasal delivery. This novel delivery platform provides a potentially groundbreaking solution, supplying a non-intrusive and direct route to the brain. This technique utilizes the inherent olfactory and trigeminal nerve pathways to bypass the BBB, allowing the administration of shRNA cargo encapsulated within common adeno-virus vectors (AAV) that specifically target brain pathways.

An Innovative Delivery Platform

In order to comprehend the mechanisms of RNA therapies better, one must initially grasp the process of genetic information flow within a cell—from DNA to RNA to protein. This core principle of biology elucidates how genetic data encoded in DNA is transcribed into RNA and subsequently acts as a blueprint for protein production. Proteins carry out an array of vital functions essential for the structure, functionality, and regulation of the cell.

RNA neuromodulators like those that utilize shRNAs work by intervening at the RNA level to regulate the flow of genetic information in the brain. shRNAs are a method employed to modulate particular genes by targeting their messenger RNA (mRNA) molecules.

The effectiveness of shRNAs in silencing a gene is reliant upon their specific sequence tailored to match the target mRNA. Upon binding with its intended target, the mRNA molecule, the shRNA prompts the breakdown of the mRNA strand, which halts the production of the protein. The precision and efficiency of shRNAs depend largely on their sequence composition. Scientists can achieve remarkable results by intricately crafting the sequence of an shRNA to perfectly align with the mRNA of the targeted gene. This meticulous design approach aims to reduce off-target effects while maximizing the therapeutic benefits. It underscores the essence of RNA therapy as a method for addressing neurological and psychiatric conditions at their molecular foundations.

Although the concept is elegantly simple, the real challenge following the design of an shRNA is enabling it to cross the BBB. The creation of an intranasal delivery platform overcomes this problem. This involves developing innovative compounds that protect shRNA from enzymatic degradation in the nasal cavity, ensuring the RNA remains intact for efficient transport across the olfactory epithelium and into the CNS. These compounds are designed to guide the shRNA to specific neuronal receptors where they can exert their therapeutic effects.

By optimizing the cargo, we see long-lasting therapeutic effects, minimizing the need for frequent administration while enhancing the overall clinical benefits. This noninvasive delivery method also improves CNS targeting by eliminating systemic side effects and maximizing the delivery of shRNA-based neuromodulators. A platform like this offers a promising avenue for treating neurodegenerative and neuropsychiatric disorders with increased efficacy and safety over traditional delivery methods.

Potential for Disease Course Reversal

Emerging research shows that targeted genetic therapies have the ability to do more than merely alleviate symptoms. These RNA neuromodulators can restore the natural brain state and gene function by suppressing abnormal neuronal expression. This approach offers the promise of slowing or even reversing illness progression at the molecular level—a paradigm shift from traditional treatments that often only manage symptoms without addressing underlying causes.

The intranasal delivery of RNA neuromodulators represents a convergence of several scientific advances: precise gene targeting, innovative delivery systems, and enhanced understanding of neurological pathways. Together, these developments are opening new frontiers in our ability to address some of the most challenging conditions affecting the brain, from neurodegenerative disorders to psychiatric conditions that have long resisted effective treatment.

As research in this area continues to advance, the prospect of delivering targeted genetic interventions directly to the brain without invasive procedures brings us closer to a new era in neurotherapeutics—one where treatment begins at the molecular level and extends to meaningful improvements in cognitive function, emotional well-being, and quality of life.

20. Comparing Methods of Gene Editing in the Design and Production of Neuromodulators

The field of genetic medicine has witnessed remarkable evolution in approaches to modifying gene expression, from DNA-based CRISPR gene editing to RNA-based neurotherapeutic applications. These distinct but complementary technologies offer different advantages for treating neurological and psychiatric conditions, each with unique mechanisms and durations of effect.

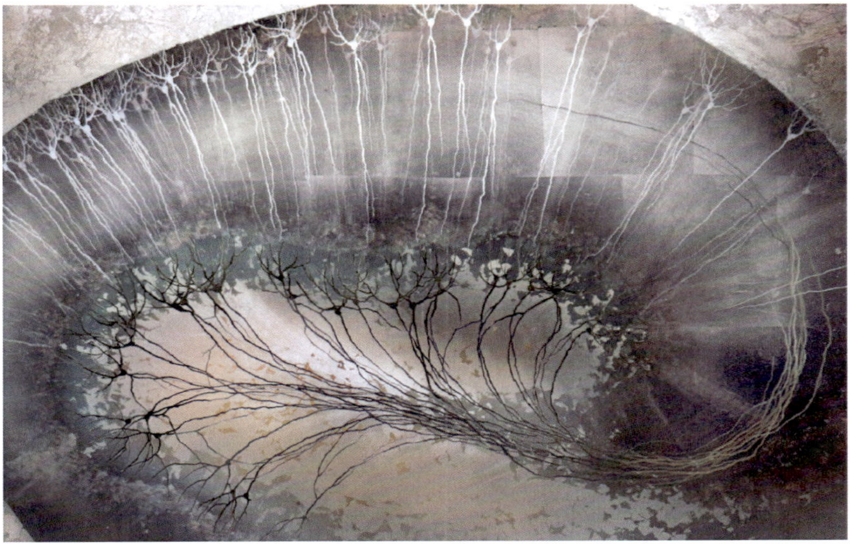

Hippocampus III, Greg Dunn

CRISPR: Permanent Genomic Editing

Researchers are employing CRISPR/Cas9 technology, a revolutionary genome-editing tool that permanently modifies DNA sequences. This approach makes it possible to directly disrupt the 5-HT2a receptor genes associated with anxiety and memory dysfunctions by inserting stop codons into the DNA, for long-term downregulation of receptor expression.

CRISPR's mechanism enables precise, irreversible modifications at the genomic level, highly effective for single-dose therapies aimed at lasting changes in neural function, particularly suitable for treating conditions with a structural genetic component, where enduring receptor modulation can yield substantial improvements in symptoms like chronic anxiety or memory impairment.

RNAi with shRNA: Temporary, Adjustable Therapeutics

Building on the success of CRISPR applications, researchers have expanded into RNA therapeutics, primarily through short hairpin RNA (shRNA). These RNA neuromodulators target the same 5-HT2a receptors but operate temporarily by inhibiting gene expression without altering the underlying DNA.

The shRNA approach to RNA interference (RNAi) enables a reversible method of silencing gene activity, thus offering flexibility for varying therapeutic needs. RNA-based treatments are more adjustable and allow infrequent dosing.

Scientists leverage intranasal delivery for both CRISPR and RNA-based compounds. This method efficiently bypasses the blood-brain barrier, enabling precise receptor targeting in brain regions linked to anxiety and memory, such as the hippocampus and limbic region. A further refinement of an shRNA molecular structure in the lab, constrains its functionality to express only within neuronal cells, adding a cell-specific precision layer to the RNAi silencing effect.

While CRISPR's DNA editing confers a lasting impact, RNA neuromodulators offer non-permanent options, making them ideal for neuropsychiatric applications, balanceing efficacy and reversibility. The transition from CRISPR-based approaches to RNA therapeutics is a shift toward patient-tailored treatments, positioning this approach at the forefront of innovative mental health therapeutics.

Comparative Analysis of CRISPR and RNA-Based Approaches

The following examines the approaches of RNA technologies (siRNA, shRNA) and CRISPR, based on their attributes, mechanisms of action, and respective advantages. This comparison reveals how each approach has distinct mechanisms, benefits, and limitations. While CRISPR offers permanent solutions, siRNA and shRNA are well-suited for temporary gene silencing.

Mechanism of Action

CRISPR: Directly edits the genome by cutting DNA at specific sites using Cas proteins and guide RNAs, for gene disruption, repair, or replacement.

siRNA: Binds to, and degrades target mRNA through the RNA-induced silencing complex (RISC), preventing translation into protein.

shRNA: Similar to siRNA but expressed from a plasmid or viral vector; it is processed in the nucleus into siRNA-like molecules that degrade target mRNA.

Targeting Specificity

- CRISPR: Highly specific; uses guide RNA to target DNA sequences directly.
- siRNA: High specificity for mRNA sequences but depends on transient presence in cells.
- shRNA: High specificity but more stable due to vector-based delivery; can have off-target effects.

Duration of Effect

- CRISPR: Permanent or long-lasting changes (requires careful design to avoid unintended edits).
- siRNA: Short-term effect lasts for days in transient transfection systems.
- shRNA: Potential long-term effect persists if the vector expresses shRNA in long-living cells.

Delivery Method

- CRISPR: Uses viral vectors, electroporation, or lipid nanoparticles to deliver Cas proteins and guide RNAs.
- siRNA: Delivered via nanoparticles, liposomes, or electroporation; not integrated into the genome.
- shRNA: Delivered via viral vectors (e.g., lentivirus or AAV); while it is unclear if it integrates into the genome, it can have stable expression in non-dividing cells.

Ease of Use

- CRISPR: Complex; requires precise design and delivery systems and involves ethical considerations.
- siRNA: Relatively simple to design and apply; transient delivery minimizes long-term risks.
- shRNA: Slightly more complex than siRNA due to plasmid/vector preparation and delivery systems.

Advantages

CRISPR

- Permanent solution for genetic conditions

- Versatile: can knock out, activate, or modify genes
- Effective for single-gene disorders and genetic engineering

siRNA

- Fast and efficient silencing of specific genes
- Non-permanent, reducing long-term risks
- No genome integration required

shRNA

- Stable and sustained gene silencing
- Ideal for long-term studies or therapeutic effects
- Efficient delivery through vectors

Challenges and Limitations

CRISPR

- Off-target effects and potential unintended mutations
- Requires advanced delivery systems
- Ethical concerns regarding germline editing

siRNA

- Transient effects may require repeated dosing
- Limited to mRNA degradation, no permanent changes
- Delivery challenges for in vivo use

shRNA

- Risk of insertional mutagenesis due to integration
- Possible immune response to viral vectors
- Requires more complex production processes

Therapeutic Applications

CRISPR

- Gene editing for monogenic disorders
- Modifying receptor-level gene expression in neurogenetics
- Precision engineering in cancer immunotherapy

siRNA

- Temporary modulation of gene expression in diseases like cancer and viral infections
- Research applications for gene function studies

shRNA

- Long-term gene silencing for chronic conditions (e.g., anxiety, cognitive disorders)
- Applications where stable, sustained effects are needed

Precision

- CRISPR: High precision, off-target edits remain a concern
- siRNA: High precision for mRNA sequences; effects reversible
- shRNA: High precision for target genes; effects are stable but not permanent

The Future of Neurogenetic Therapeutics

The comparison between CRISPR and RNA-based approaches highlights a critical evolution in neurogenetic medicine. While CRISPR offers the potential for permanent correction of genetic disorders, RNA neuromodulators provide a more nuanced, adjustable approach that may be particularly well-suited to complex neuropsychiatric conditions where fine-tuning gene expression is preferable to permanent modification.

As our understanding of both technologies advances, we're likely to see increasing specialization in their applications, with CRISPR perhaps finding its greatest utility in congenital neurodevelopmental disorders, while RNA neuromodulators become the preferred approach for acquired or progressive neuropsychiatric conditions.

The flexibility, reversibility, and cell-type specificity of RNA neuromodulators make them particularly promising for addressing conditions like anxiety, depression, and cognitive decline—disorders where subtle modulation of neural circuits may be more appropriate than permanent genetic changes. This approach aligns with the growing emphasis on personalized medicine, where treatments are increasingly tailored to individual patients' specific needs and genetic profiles.

The future of neurotherapeutics likely lies in the thoughtful integration of both approaches, with treatment selection guided by the nature of the condition, the genetic mechanisms involved, and the desired duration and reversibility of effect. This personalized approach to genetic medicine represents one of the most promising frontiers in addressing previously intractable neurological and psychiatric disorders.

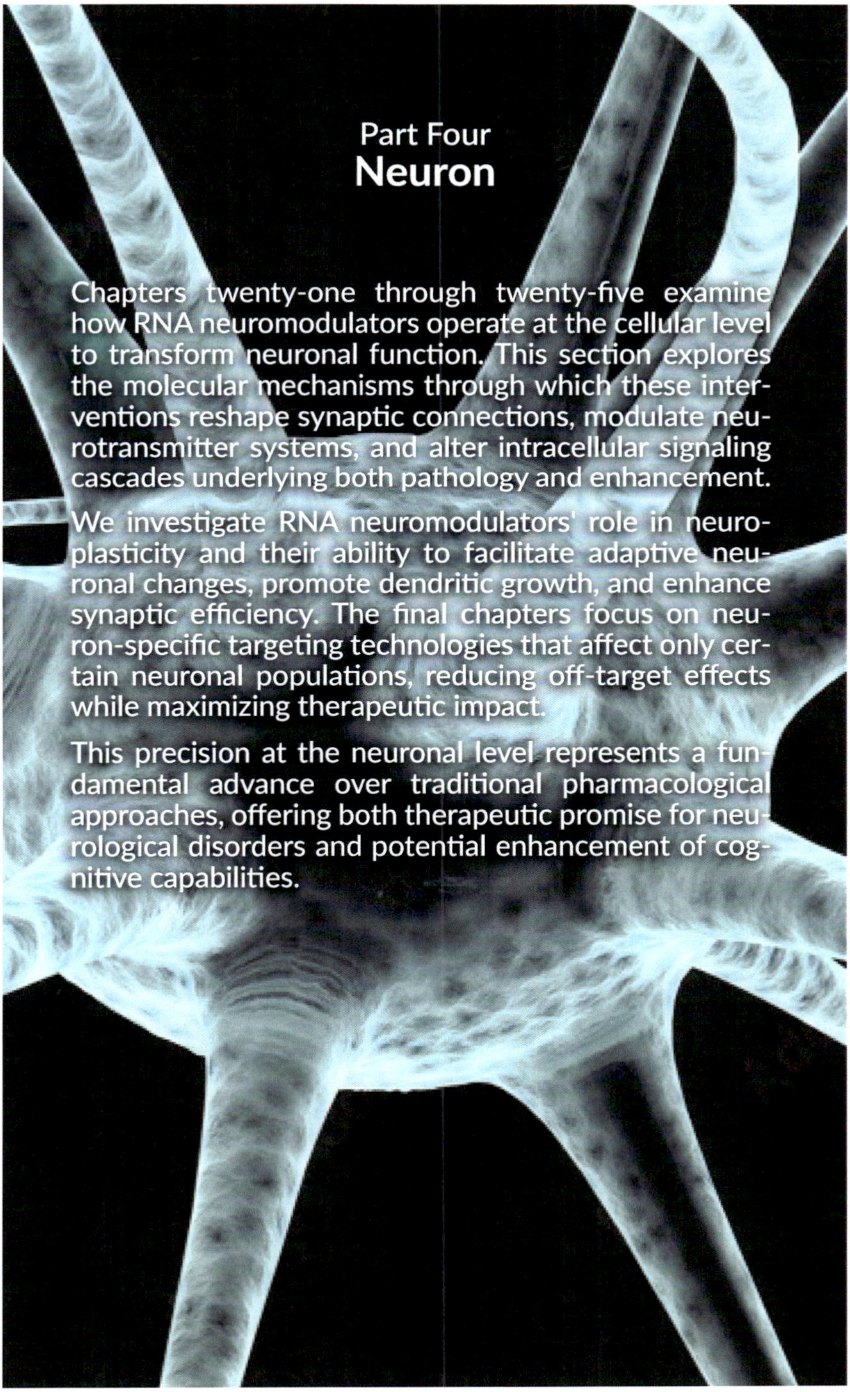

Part Four
Neuron

Chapters twenty-one through twenty-five examine how RNA neuromodulators operate at the cellular level to transform neuronal function. This section explores the molecular mechanisms through which these interventions reshape synaptic connections, modulate neurotransmitter systems, and alter intracellular signaling cascades underlying both pathology and enhancement.

We investigate RNA neuromodulators' role in neuroplasticity and their ability to facilitate adaptive neuronal changes, promote dendritic growth, and enhance synaptic efficiency. The final chapters focus on neuron-specific targeting technologies that affect only certain neuronal populations, reducing off-target effects while maximizing therapeutic impact.

This precision at the neuronal level represents a fundamental advance over traditional pharmacological approaches, offering both therapeutic promise for neurological disorders and potential enhancement of cognitive capabilities.

21. Precise Genetic Targeting in Neuropsychiatric Disorders: Precise, Effective, and Safer Care

Recent advancements in RNA-based therapies, such as small interfering RNAs (siRNAs) and short hairpin RNAs (shRNAs), offer promising tools for noninvasive gene modulation. Chemical modifications have improved the stability and pharmacokinetics of siRNAs, enabling them to cross the blood-brain barrier (BBB) and reach target brain regions.

Motor-Parietal Cortex, Greg Dunn

The BBB is a protective barrier of tightly packed cells that separates the brain from the bloodstream, preventing harmful substances from entering the brain while allowing essential nutrients and gases to pass through. It plays a crucial role in maintaining the brain's stable environment. This natural defense system, while vital for normal brain function, has historically presented a significant challenge for delivering therapeutics to treat neurological disorders.

RNA neuromodulators are delivered through viral vectors that transport genetic material directly into neurons for specific gene expression control. This approach is highlighted as noninvasive and effective in bypassing the blood-brain barrier to target specific neural circuits and can specifically modulate neuronal receptors linked to mental disorders. These innovations, combined with delivery methods like the nose-to-brain route, enhance RNA therapeutics' precision and effectiveness for treating neurological diseases.

RNA neuromodulators that regulate proteins associated with synaptic dysfunction, such as those contributing to hyperactivity, can help restore normal neurotransmission and brain plasticity. By targeting the molecular and neural network mechanisms underlying these dysfunctions, this approach offers a more upstream therapeutic strategy for treating cognitive impairment, neurodegenerative diseases, schizophrenia, and anxiety disorders. Addressing these core drivers holds the potential for more effective interventions by directly modulating the root causes of synaptic and network dysfunction.

Sustainable Treatment Solutions

Research suggests that genetic therapies have the potential to treat neurocognitive and neuropsychiatric disorders by addressing their underlying causes at the molecular level. These therapies offer long-lasting benefits, reducing the need for continuous medication and shifting the paradigm in mental health treatment. By directly targeting the molecular triggers of these conditions, genetic editing techniques provide a more precise and durable approach to managing and potentially reversing such illnesses, offering a transformative framework for mental health care.

The sustainability of RNA neuromodulators lies in their ability to create lasting changes in gene expression with minimal intervention. Unlike traditional pharmaceuticals that require daily dosing to maintain therapeutic levels, a single administration of these neuromodulators can potentially provide months of therapeutic benefit. This extended duration not only improves patient adherence but also reduces the cumulative burden of side effects that often accompanies chronic medication use.

Pre-Clinical Findings

Rodent model studies have shown significant promise for RNA neuromodulators in precisely targeting molecular pathways involved in neuropsychiatric disorders. These therapies demonstrate the ability to downregulate specific neuronal receptors, reducing symptoms and improving brain function. Preclinical findings highlight the potential of targeted genetic interventions to address critical pathways involved in cognitive functions like memory, signaling a new era in precision psychiatric treatment.

The specificity of these interventions represents a fundamental advance over traditional approaches. By modulating only the precise molecular targets associated with disease pathology, RNA neuromodulators can potentially avoid the widespread effects on multiple receptor systems that characterize most current psychiatric medications. This precision not only enhances efficacy but significantly reduces unwanted side effects.

As research in this field continues to advance, the convergence of genetic precision, innovative delivery methods, and enhanced understanding of neuropsychiatric disease mechanisms is creating unprecedented opportunities for transforming mental health care. These emerging technologies offer hope for developing treatments that address the fundamental causes of brain disorders rather than merely managing their symptoms, potentially changing the trajectory of conditions that have long resisted effective intervention.

22. How RNA Therapies Target Neurons to Treat Neurological Disorders

RNA interference (RNAi) and short hairpin RNA (shRNA) can target particular sites on specific neuron cells precisely. With that property, treating neurological disorders directly becomes possible. Let's analyze how RNA therapies target specific neurons.

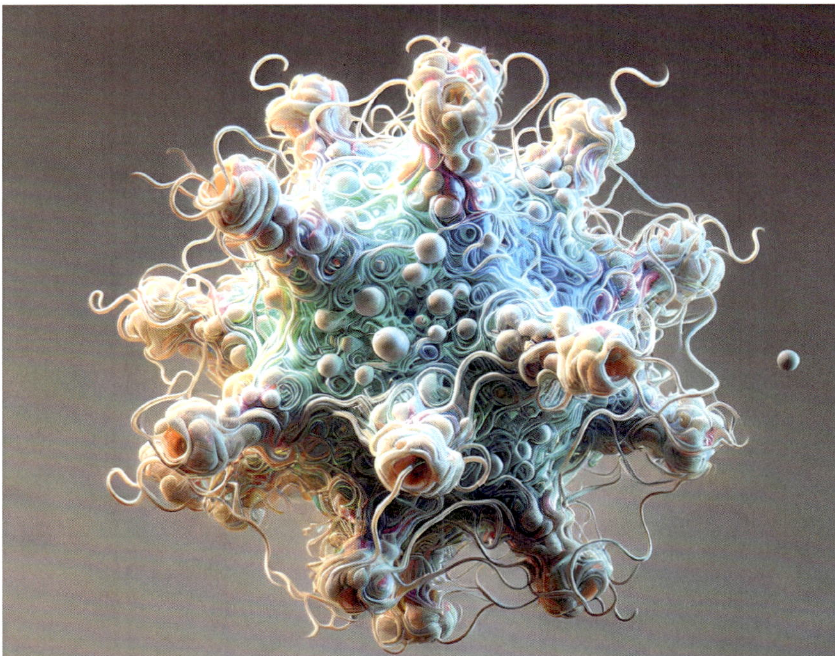

5-HT2a Neuronal Receptor. Rendering.

RNAi Mechanism Targeting Neurons

RNA interference (RNAi) is a natural biological process found in all cells, including neurons. This system can be harnessed for therapeutic purposes to selectively modulate gene expression using small RNA molecules such as short hairpin RNA (shRNA) or small interfering RNA (siRNA).

When shRNA is introduced into a neuron, it undergoes the following process:

1. *Conversion:* The cellular enzyme Dicer converts shRNA into siRNA.
2. *RISC Formation:* The siRNA is incorporated into the RNA-induced silencing complex (RISC), a multi-protein complex that includes the Argonaute protein.
3. *Target Recognition:* The RISC uses the siRNA as a guide to locate and bind to complementary sequences on specific messenger RNA (mRNA) molecules.
4. *mRNA Degradation:* Once bound, the RISC cleaves the target mRNA or inhibits its translation, effectively "silencing" the gene by preventing protein production.

This process takes advantage of a natural cellular pathway that regulates protein levels. Normally, mRNA is single-stranded when translated into protein. However, the presence of double-stranded RNA (as created by the binding of siRNA to mRNA) triggers rapid degradation, a mechanism cells use to control protein concentrations based on their needs.

In therapeutic applications, this system can be used to deliberately decrease the concentration of particular proteins. For example:

1. *Target:* 5-HT2a receptor in neurons
2. *Method:* Introduction of shRNA designed to recognize the mRNA for the 5-HT2a receptor
3. *Result:* Decreased expression of the receptor
4. *Potential Effects:* Altered neuronal excitability, potentially enhancing memory and reducing anxiety

By selectively targeting specific genes, this technique allows for precise manipulation of neuronal function at the molecular level. This approach offers promising possibilities for treating various neurological and psychiatric conditions by modulating the expression of key proteins involved in these disorders.

Techniques for Targeted Delivery of Substances to Specific Neurons

Vectors are vehicles used to deliver genetic material into target cells for gene editing purposes. They are essential tools in genetic engineering, gene therapy, and molecular biology research. Gene therapy often utilizes vectors to transport genetic material into cells effectively and safely as part of the treatment process for various conditions. The use of adeno-associated viruses (AAVs), a type of viral vector known for its efficiency and safety in gene delivery methods, is quite common due to their ability to transfer therapeutic RNA directly into neurons with high efficacy. AAVs' reputation for not causing harm to humans adds to their appeal for applications involving genetic treatments. Viruses have developed naturally to effectively invade cells and transport their material efficiently. This natural ability is utilized by scientists who alter these viruses to transport shRNA instead of their original viral content.

They transform them into carriers that can pinpoint and target cells, like neurons, in the brain. The unique nature of AAV vectors is boosted by employing promoters—DNA segments that dictate the timing and location of gene activity in the body. By utilizing promoters that are active solely in specific neuron types, researchers can fine-tune the AAV vectors to transport genetic material exclusively to the intended cell types in the brain. This precise targeting reduces the chances of consequences, such as therapeutic RNA impacting unintended cells and causing possible side effects.

Treating disorders poses a significant hurdle when it comes to delivering medications through the BBB. Scientists have found ways to modify AAVs to bypass the BBB by using existing transportation systems that help move substances across this protective barrier in the brain efficiently. This breakthrough technique enables the administration of therapeutic drugs directly to targeted areas in the brain, without invasive procedures that are challenging otherwise.

The intranasal route of administration circumvents the blood-brain barrier, promoting direct entry into the brain. Through precise delivery, RNA drugs may be directly administered to brain areas accessible via the olfactory pathways, allowing for efficient targeting of neurons.

Directly delivering drugs into the brain or cerebrospinal fluid is also possible. These treatments can be quite invasive and pose serious risks. They require creation of a cranial aperture to administer medicines. Despite their invasive nature, these therapies are pursued because of a profound level of suffering experienced by individuals, driving them to explore experimental interventions.

Neurological Disorders Management

RNA therapies, particularly those targeting the 5-HT2a receptor, hold significant promise in managing neurological disorders by precisely targeting specific brain circuits. With Alzheimer's disease, targeting 5-HT2a receptors may improve memory impairments and alleviate related conditions like anxiety and depression. Further research is needed to confirm these benefits, but early evidence suggests that RNA therapies could lead to better patient outcomes.

In epilepsy, downregulating the excitatory 5-HT2a receptor using RNA interference, such as shRNA, could reduce the frequency and severity of seizures by directly modulating neuronal activity. Similarly, in schizophrenia, where current treatments block the 5-HT2a receptor but often cause side effects, precision RNA therapies offer a more targeted and potentially safer alternative. By specifically downregulating the 5-HT2a receptor, shRNA therapies could alleviate many symptoms of schizophrenia and other serotonin-related conditions without the broad, systemic effects of traditional medications.

The advancement of RNA interference therapies, such as shRNA, in targeting the 5-HT2a receptor, presents a promising approach for treating a range of neurological disorders. This method allows for precise modulation of neuronal circuits, offering a safer and more effective treatment option compared to existing drugs, potentially improving patient well-being and quality of life.

Improving Cognitive Abilities in People Without Health Conditions

RNA therapies not only offer potential treatments for medical conditions but also show promise in enhancing cognitive function in healthy individuals by improving memory and learning. Evidence from animal studies demonstrates that targeting specific genes involved in synaptic plasticity and neurogenesis can significantly enhance cognitive abilities. By focusing on genes linked to stress responses, RNA therapies can also reduce anxiety and improve overall mental well-being through precise receptor modulation.

Challenges and Opportunities in RNA Therapy

Maximizing the effectiveness of RNA therapies requires targeting specific neurons and ensuring broad access to these treatments once they are approved for clinical use. RNA-based therapies precisely target specific neuronal pathways, providing a new approach to treating neurological disorders and improving cognitive performance. These therapies have the potential to revolutionize mental health treatment by addressing the underlying neural mechanisms.

23. Harnessing RNA Therapies to Address Neuroinflammation

Researchers are developing RNA-based medications that specifically and efficiently target neuroinflammatory pathways, offering a new approach to treating conditions like cognitive impairment and anxiety spectrum disorders.

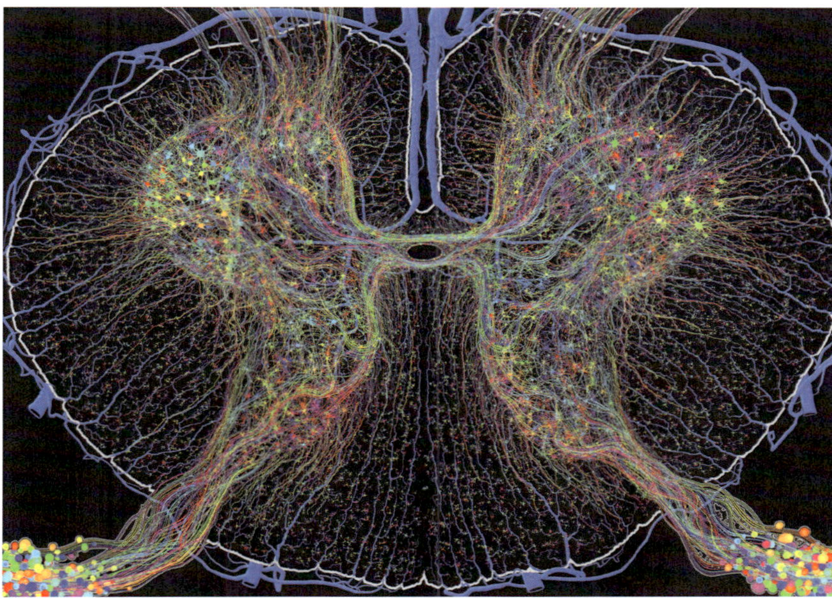

Grey Matter Glia and Blood Vessals, **Greg Dunn**

These RNA therapeutics aim to reduce symptoms by targeting neural networks and receptors associated with mental health conditions. Short hairpin RNA (shRNA) plays a crucial role in these treatments, which can be delivered intranasally to target specific regions of the brain, bypassing the blood-brain barrier and minimizing side effects.

RNA-based therapies have become the new methods for directly targeting neuroinflammation, a crucial factor in several neurological and psychiatric disorders. A primary factor contributing to the effectiveness of RNA therapies is their capacity to alter gene expression precisely. They target the fundamental molecular origins of disorders. Here, we investigate novel methodologies and lay the foundation for groundbreaking solutions.

Accurate Targeting

A substantial obstacle must be overcome to transport RNA molecules across the blood-brain barrier (BBB). Conversely, progress in several transportation methods, such as lipid nanoparticles (LNPs), exosomes, and viral vectors like adeno-associated viruses, simplifies how they get there. LNPs demonstrate effective delivery of mRNA vaccines. They encapsulate RNA therapeutic compounds, preventing breakdown. They also enhance their absorption by targeting the specific neuronal receptors on particular cells. Researchers are also working with AAVs that are modified to pass through the blood-brain barrier effectively. These modified AAVs are intended to deliver RNA therapies directly to neurons. The actual use of this concept demonstrates its ability to provide a focused and accurate method for treating brain disorders.

Determinants of the Immune Response

RNA molecules are inherently unstable. They can elicit responses from the immune system, hindering their efficacy as medicinal agents. To address this problem, scientists use chemical alterations such as 2'-O-methylation and phosphorothioate backbones. These alterations improve the stability of the RNA molecules and decrease their capacity to provoke an immune response, making them more appropriate for therapeutic applications. It is vital to ensure that the therapeutic RNA remains effective when dealing with conditions that require therapy over a lengthy period.

Expanding the Range of Therapeutic Targets

RNA treatments have effectively treated various diseases unrelated to a particular gene. They can tackle complex disorders that encompass several genes and pathways. RNA therapeutics can target many essential pathways in neuroinflammatory illnesses simultaneously. Alzheimer's and Parkinson's both involve multiple molecular operations. Using a multi-target approach may slow down the progression of the disease and improve overall therapy outcomes. Other scientists are currently focusing on regulating gene expression in specific genes.

The advancement of genetic profiling and sequencing technology has facilitated the emergence of tailored medicine and expanded the scope of research on RNA therapeutics. Understanding an individual's genetic composition may customize treatment for specific conditions. The compounds target genetic variants and mutations linked to neuroinflammation. These conditions may require a personalized approach to improve effectiveness.

Future Developments and Advancements

There are exciting prospects for integrating RNA therapies with digital health technology, such as wearable devices and AI-driven analytics, in the future. These technologies will provide real-time monitoring of pharmacological responses and the progression of illnesses, enabling more precise adjustments to therapeutic regimens. Our knowledge of the brain's molecular composition is progressing, and mapping neural networks is expanding our understanding of specific brain functions. New RNA-based treatments focusing on previously unknown endpoints will arise.

The initial stages of developing RNA-based treatments for neuroinflammation are now underway. Ongoing research and technological advancements are leading to the emergence of a promising era of more precise medical treatments. These therapies may significantly change the current choices for managing various neurological and mental disorders. We expect this to result in a reduction in distress and an enhancement of the patient's overall well-being.

24. Synthetic Biology: A New Era in Biotechnology

By J. L. Mee

Synthetic biology represents the future, built by the most potent and sustainable manufacturing platform humanity has ever had. We are on the cusp of a breathtaking new industrial evolution.

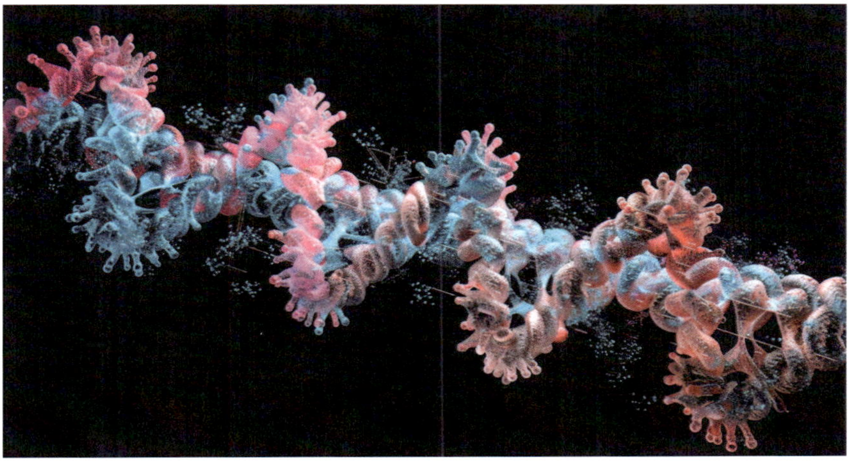

Protein in posttranslational modification. Rendering.

Synthetic biology already exists, and functions in its early stages, but the supporting businesses and ancillary players are yet to be fully established. In the future, synthetic biology will become a general-purpose technology with vast societal value, similar to the phone and the internet.

A May 2020 McKinsey study analyzed four hundred synthetic-biology-related innovations and found that these could generate an average of $4 trillion annually until 2040. This figure doesn't include the related businesses, services, and products that will emerge to support the industry. This development of interconnected companies is known as a value network. For example, the Internet of Things includes software, platforms, interfaces, connectivity, security, agriculture, health care, vehicles, supply chains, robotics, industrial wearables, and many other subcategories. Synthetic biology's value network is just beginning, with investment growth doubling yearly since 2018. This interest supports companies that manufacture synthesizing machines, robots, and assemblers; sell DNA, enzymes, proteins, and cells; and develop specialized software tools. We are entering a new era: the Biological Age, where manipulating molecules, engineering microorganisms, and building biocomputing systems will unlock new business opportunities, mitigate environmental damage, and improve the human condition.

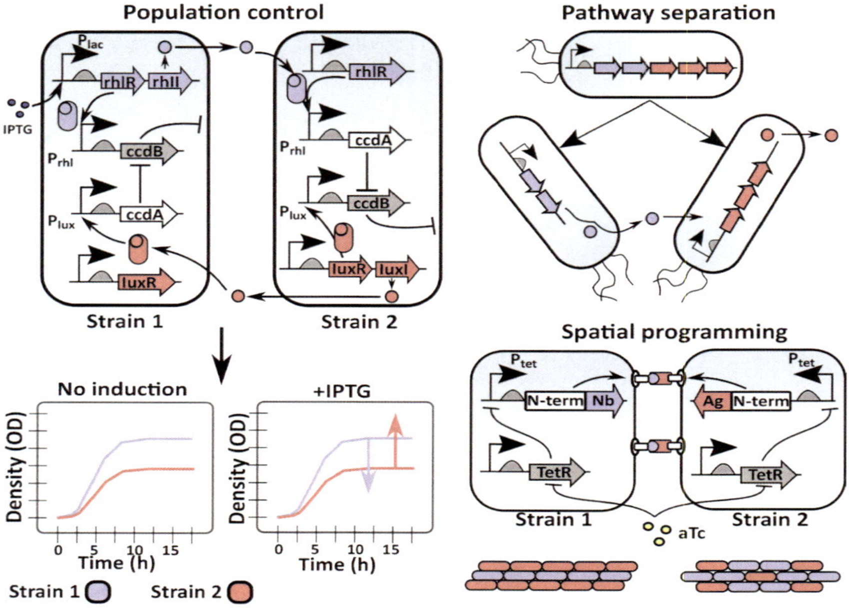

Engineering Behaviors of Synthetic Microbial Consortia:
Nicholas S. McCarty and Rodrigo Ledesma-Amaro, *Trends in Biology*

Synthetic biology can bring personalized health care, address food insecurity, create safer industrial manufacturing and agriculture, tackle climate change, and even enable off-planet living. However, these advancements also raise serious questions about equity, ethics, geopolitical risk, and national security. The implications of manipulating life are profound, potentially influencing societies, economies, national security, and geopolitical alliances in unimaginable ways.

The future of life, potentially shaped by a select group of decision-makers with specialized skills and knowledge, raises critical questions about the ethical use of synthetic biology. The potential for social inequality, if only the wealthy can afford these advancements, underscores the need for equitable access to this technology. Despite inevitable divisions due to differing levels of trust in science and access to tools, we must strive to ensure that synthetic biology is accessible to all.

Short-term thinking in the private sector has shown its damaging effects, such as prioritizing profits over societal well-being. Evaluating synthetic biology's impact requires considering wealth distribution, job markets, privacy attitudes, socioeconomic factors, political dynamics in China, the EU, and the US, and space initiatives like terraforming Mars. Synthetic biology intersects with other tech areas, including AI, telecommunications, blockchain, consumer electronics, social media, robotics, and algorithmic surveillance, all increasingly vital in the bio-economy.

Businesses should prepare for disruption as synthetic biology intersects with every industry sector, leading to a remarkable transformation of our societies and species.

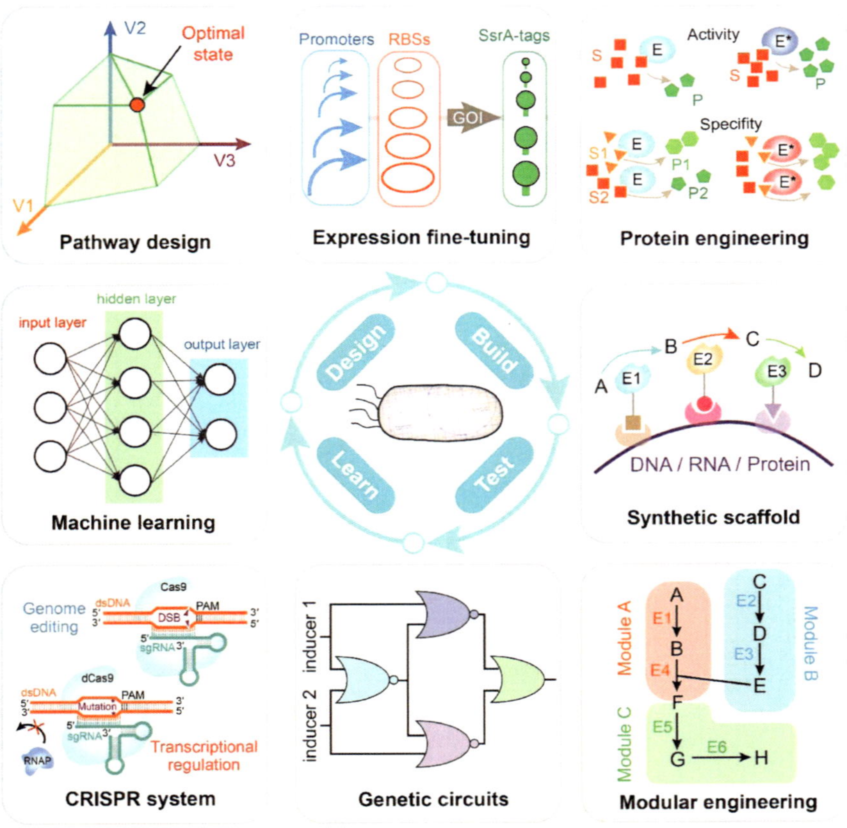

Synthetic biology for future food, Xueqin Lv, et al.

25. Wide-Ranging Benefits of 5-HT2a Receptor Reduction

The reduction of 5-HT2a post-synaptic receptor density and activity represents one of the most promising targets for RNA therapeutics in mental health treatment. This approach has been linked to neuropsychiatric stability, cognitive improvements, and neuroprotection through its effects on key neuromodulator systems (Celada et al., 2013; Weisstaub et al., 2006). The wide-ranging benefits span multiple domains of brain function and overall health, offering hope for conditions that have proven resistant to conventional treatments (Nichols, 2016).

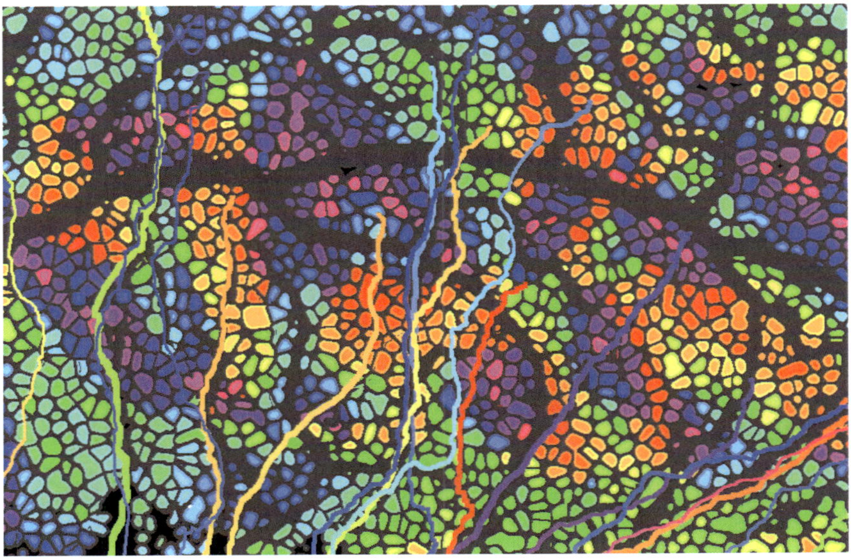

Gray Matter, Detail, Greg Dunn

The growing body of research into 5-HT2a receptor modulation suggests why this particular neuromodulator system has become such a focal point for next-generation therapeutics. As we explore the evidence, the potential for targeted interventions becomes increasingly compelling.

Below, I've organized the potential benefits of 5-HT2a receptor reduction by category, followed by key supporting research. This comprehensive overview demonstrates why RNA therapeutics targeting this receptor system could represent a paradigm shift in mental health treatment.

I. Cognitive Enhancement & Memory Improvement

Reducing 5-HT2a receptor activity offers potential cognitive benefits through several mechanisms:

May reduce hippocampal hyperactivity, potentially improving memory consolidation and recall by allowing more efficient information processing

May enhance cognitive flexibility by modulating thought patterns that could interfere with adaptive thinking

Appears to support executive function, potentially aiding in problem-solving and decision-making through effects on prefrontal cortex function

May promote synaptic plasticity, potentially improving learning and adaptability through neuromodulator interactions

Could prevent memory distortions associated with 5-HT2a overactivation, potentially leading to more accurate recall

Research examining cognitive outcomes suggests that modulation of these receptors may influence cognitive processing networks (Preller et al., 2019), potentially offering new approaches for addressing cognitive impairment where traditional medications have shown limited efficacy (Geyer & Vollenweider, 2008).

Key Research

1. Celada, P., Puig, M. V., & Artigas, F. (2013). Serotonin modulation of cortical neurons and networks. *Frontiers in Integrative Neuroscience*, 7, 25.

2. Geyer, M. A., & Vollenweider, F. X. (2008). Serotonin research: Contributions to understanding psychoses. *Schizophrenia Bulletin*, 34(1), 30-43.

3. Preller, K. H., et al. (2019). The fabric of meaning and subjective effects in LSD-induced states depend on serotonin 2A receptor activation. *Current Biology*, 29(21), 3912-3920.

II. Anxiety & Stress Reduction

The relationship between 5-HT2a receptors and anxiety has been investigated in several studies:

May decrease fear responses, potentially reducing anxiety by modulating amygdala activity.

Could enhance emotional resilience, possibly reducing susceptibility to chronic stress through neuromodulator regulation.

May reduce amygdala hyperactivity, potentially preventing overreaction to stressors through receptor-mediated effects.

Could help regulate stress responses, potentially reducing the risk of stress-related symptoms.

May influence ruminative thinking and intrusive thoughts, potentially affecting obsessive-compulsive tendencies.

Studies have demonstrated that anxiety-related behaviors can be modulated by 5-HT2a receptor activity (Weisstaub et al., 2006). This relationship appears to involve structures associated with fear processing (Zuo et al., 2021), suggesting that targeted RNA therapeutics affecting this receptor system might provide a novel approach for anxiety disorders.

Key Research

1. Weisstaub, N. V., et al. (2006). Cortical 5-HT2a receptor signaling modulates anxiety-like behaviors in mice. *Science*, 313(5786), 536-540.

2. Zuo, Y., et al. (2021). The role of serotonin 5-HT2a receptors in stress and resilience. *Neuroscience & Biobehavioral Reviews*, 124, 216-231.

3. Nichols, D. E. (2016). Psychedelics. *Pharmacological Reviews, 68(2), 264-355.*

III. Mood Stabilization & Psychiatric Benefits

The impact of 5-HT2a receptor modulation has been studied in relation to psychiatric conditions:

May affect psychotic symptoms, including hallucinations and delusions that are targets in schizophrenia treatments.

Could influence mood stability in bipolar disorder, potentially affecting cycling between mood states.

May have effects on depression symptoms, particularly in treatment-resistant cases through novel mechanisms.

Might reduce drug-induced hallucinatory states, potentially beneficial for individuals with sensitivity to such effects.

Could affect emotional regulation, potentially influencing impulsivity and mood fluctuations through neuromodulator systems.

Neuroimaging studies have documented alterations in 5-HT2a receptor distribution in patients with mood disorders (Ettrup et al., 2016). Some research suggests that modulation of these receptors may affect network-level brain activity in treatment-resistant depression (Carhart-Harris et al., 2017), raising the possibility that RNA therapeutics targeting this system could potentially address aspects of the underlying dysregulation. This approach may warrant investigation for patients who have not responded adequately to conventional treatments (Vollenweider & Kometer, 2010).

Key Research:

1. Carhart-Harris, R. L., et al. (2017). Psilocybin for treatment-resistant depression: fMRI-measured brain network changes. *Scientific Reports,* 7, 13187.

2. Ettrup, A., et al. (2016). Serotonin 2a receptor agonist binding in the human brain: A PET study. *NeuroImage,* 130, 184-191.

3. Vollenweider, F. X., & Kometer, M. (2010). The neurobiology of psychedelic drugs: Implications for mood disorders. *Nature Reviews Neuroscience,* 11(9), 642-651.

IV. Neuroprotective & Anti-Aging Benefits

Research suggests potential neuroprotective effects of 5-HT2a receptor modulation:

May influence neuroinflammatory processes, which could be relevant to neurodegenerative diseases.

Could have effects on excitotoxicity mechanisms, potentially affecting neuronal viability.

May be relevant to cognitive aging, particularly in relation to neurodegenerative processes.

Could affect brain network organization, potentially influencing information processing.

May have effects on cellular function, potentially relevant to neural health.

Research has suggested potential connections between serotonergic systems and aging-related processes (Martin et al., 2014, not Amaral & Sinnamon, 1977). Studies indicate that 5-HT2a receptors may influence cellular processes related to neuronal health (Zhang & Stackman, 2014), suggesting that targeted RNA therapeutics could potentially address certain aspects of neurodegeneration. This area represents an emerging field of investigation that may contribute to our understanding of neurodegenerative conditions and potential therapeutic approaches (Passie et al., 2008).

Key Research References for Neuroprotection

1. Zhang, G., & Stackman, R. W. (2014). The role of serotonin 5-HT2a receptors in memory and cognition. *Brain Research Bulletin,* 104, 81-89.

2. Martin, D. A., Marona-Lewicka, D., Nichols, D. E., & Nichols, C. D. (2014). Chronic LSD alters gene expression profiles in the mPFC relevant to schizophrenia. *Neuropharmacology,* 83, 1-8.

3. Passie, T., et al. (2008). The pharmacology of psilocybin. *Addiction Biology,* 13(2), 229-238.

V. Sleep & Circadian Rhythm Regulation

Sleep processes and 5-HT2a receptors appear to have important relationships:

May influence slow-wave sleep patterns, which are important for memory consolidation.

Could affect sleep quality parameters, potentially reducing certain insomnia symptoms.

May play a role in circadian rhythm regulation, potentially affecting sleep-wake cycles.

Could influence certain types of sleep-related perceptual experiences, such as hypnagogic phenomena.

Studies have identified that 5-HT2a receptors may influence sleep regulation mechanisms (Monti, 2011). Research suggests these receptors may affect sleep architecture through pathways distinct from those targeted by conventional sleep medications (Landolt, 2012), potentially offering different approaches to treating certain sleep disturbances associated with psychiatric conditions (Sharpley et al., 1994). This suggests that RNA therapeutics targeting 5-HT2a receptors could potentially address some aspects of sleep disruption through novel mechanisms.

Key Research:

1. Monti, J. M. (2011). Serotonin control of sleep-wake behavior. *Sleep Medicine Reviews,* 15(4), 269-281.

2. Landolt, H. P. (2012). Sleep homeostasis: A role for serotonin 5-HT2a receptors? *Neuropsychopharmacology,* 37(2), 261-263.

3. Sharpley, A. L., et al. (1994). The effects of paroxetine and other SSRIs on sleep: Evidence for drug-specific effects. *Neuropsychobiology,* 29(3), 174-178.

VI. Sensory Processing & Perceptual Stability

Research has examined 5-HT2a receptor functions in sensory processing:

May influence sensory filtering mechanisms, potentially affecting sensitivity to sensory stimuli.

Could play a role in attention regulation, potentially affecting focus and distractibility.

May affect perceptual processing, potentially relevant to certain perceptual alterations.

Could influence sensorimotor integration, potentially affecting *movement coordination.*

May be involved in sensory aspects of anxiety, potentially relevant to sensory-related anxiety responses.

Research has established that 5-HT2a receptors are involved in aspects of sensory perception and filtering processes (Geyer, 1998). Studies have identified abnormal activity of these receptors in conditions with perceptual disturbances (Vollenweider, 2001). Brain imaging research has shown that modulation of 5-HT2a receptor activity can affect patterns of connectivity associated with perceptual processing (Kometer et al., 2013), suggesting that RNA therapeutics targeting this system might potentially benefit individuals with certain types of sensory processing alterations.

Key Research

1. Geyer, M. A. (1998). The role of serotonin 5-HT2a receptors in sensory processing. *Neuroscience & Biobehavioral Reviews,* 23(5), 601-613.

2. Kometer, M., et al. (2013). Psilocybin-induced changes in global brain connectivity correlate with hallucinatory states. *Biological Psychiatry,* 73(8), 731-738.

3. Vollenweider, F. X. (2001). Positron emission tomography and serotonergic modulation of perception and cognition in schizophrenia. *Schizophrenia Research,* 50(1-2), 79-88.

VII. Addiction & Impulse Control

Research has examined connections between 5-HT2a receptors and addiction-related behaviors:

May influence compulsive behavior patterns, potentially relevant to addiction treatment approaches.

Could affect aspects of impulse regulation, potentially impacting decision-making processes.

May be involved in behavioral conditioning, potentially relevant to therapeutic interventions.

Could interact with dopaminergic systems, which are implicated in reward and addiction.

May affect emotional regulation mechanisms potentially relevant to impulsivity-related behaviors.

Studies have identified connections between 5-HT2a receptor function and addiction-related behaviors (Howell & Cunningham, 2015). In preclinical research modulation of these receptors affects impulsivity in experimental models (Barlow et al., 2015), suggesting potential mechanisms by which RNA therapeutics targeting this system might influence impulse control. This area of investigation may be relevant for developing new approaches to substance use disorders. (Nichols, 2018).

Key Research

1. Howell, L. L., & Cunningham, K. A. (2015). Serotonin 5-HT2a receptors in drug addiction. *Molecular Psychiatry*, 20(2), 182-191.

2. Nichols, D. E. (2018). Psychedelics as therapeutics in addiction treatment. *Neuropharmacology*, 142, 83-91.

3. Barlow, R. L., et al. (2015). 5-HT2a receptor blockade reduces impulsivity in addiction-prone rats. **Psychopharmacology**, 232(1), 129-140.

VIII. Cardiovascular & Metabolic Benefits

Emerging research suggests potential relationships between 5-HT2a receptors and physiological functions:

May be involved in blood pressure regulation mechanisms, potentially relevant to stress-related cardiovascular effects.

Could affect aspects of glucose metabolism, potentially influencing metabolic processes.

May interact with stress hormone pathways, potentially relevant to metabolic regulation.

Could influence vascular function, potentially affecting cerebral blood flow.

May be involved in neuroendocrine regulation, potentially relevant to metabolic health.

Research identified that 5-HT2a receptors influence cardiovascular function (Mück-Seler et al., 2002) and involved in aspects of metabolic regulation (Watanabe et al., 2011). These receptors may be part of pathways that affect neuroendocrine functions, raising the possibility that RNA therapeutics targeting 5-HT2a receptors might potentially address certain aspects of stress-related physiological dysregulation (Beindorff et al., 2018).

Key Research

1. Mück-Seler, D., et al. (2002). Serotonin and blood pressure regulation: Role of 5-HT2a receptors. *Journal of Hypertension*, 20(2), 253-259.

2. Watanabe, H., et al. (2011). The impact of serotonin on metabolic syndrome. *Endocrine Journal**, 58(1), 1-11.

3. Beindorff, N., et al. (2018). Serotonin 5-HT2a receptors in metabolic regulation. *Trends in Endocrinology & Metabolism*, 29(9), 625-635.

Conclusion: A Promising Target for RNA Therapeutics

The diverse functions of 5-HT2a receptors across multiple biological systems make them a potentially valuable target for RNA therapeutic approaches. By modulating this receptor system, there may be opportunities to address multiple aspects of neuropsychiatric disorders through these mechanisms.

The potential advantage of RNA therapeutics in targeting 5-HT2a receptors lies in their ability to achieve more precise control of gene expression, potentially allowing for more selective modulation of receptor function in specific brain regions.

The scientific evidence regarding 5-HT2a receptor modulation continues to develop, suggesting that RNA therapeutics targeting this neuromodulator system could represent a novel direction in mental health treatment research—one that may address certain underlying mechanisms rather than merely managing symptoms. As investigation in this field progresses, our understanding of how these approaches might benefit patients with various conditions will continue to evolve.

Research into RNA therapeutics targeting 5-HT2a receptors represents an important frontier in the development of new treatment approaches for neuropsychiatric conditions. While much work remains to be done to establish efficacy and safety, the potential to address conditions that have not responded adequately to current treatments makes this a valuable area for continued investigation.

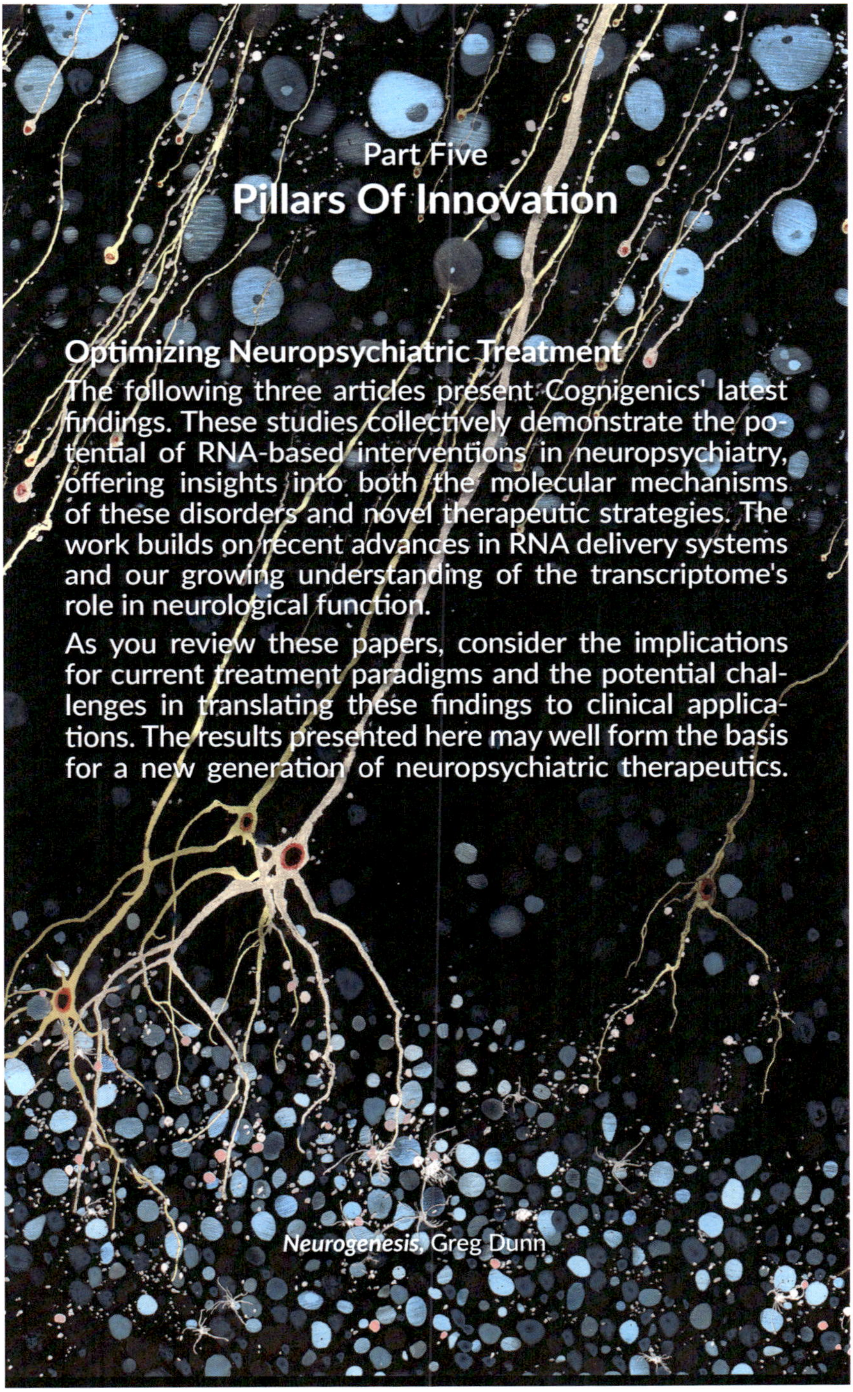

Part Five

Pillars Of Innovation

Optimizing Neuropsychiatric Treatment

The following three articles present Cognigenics' latest findings. These studies collectively demonstrate the potential of RNA-based interventions in neuropsychiatry, offering insights into both the molecular mechanisms of these disorders and novel therapeutic strategies. The work builds on recent advances in RNA delivery systems and our growing understanding of the transcriptome's role in neurological function.

As you review these papers, consider the implications for current treatment paradigms and the potential challenges in translating these findings to clinical applications. The results presented here may well form the basis for a new generation of neuropsychiatric therapeutics.

Neurogenesis, Greg Dunn

26. Scientific Review

The following three academic papers, published in top-tier scientific journals, form a foundation–the pillars of innovation for RNA therapeutics, treating individuals with neuropsychiatric and neurodegenerative conditions. These papers provide the critical scientific basis for groundbreaking therapeutic advancements shaping this field.

The emerging field of RNA-based therapeutics marks a pivotal shift in how we address neuropsychiatric and neurodegenerative disorders. Historically, treatments for conditions such as anxiety, mild cognitive impairment, and early-onset Alzheimer's disease have been limited to symptomatic management. The advent of precision genetic modulation offers a potentially transformative approach by targeting the underlying molecular and neuronal mechanisms of these disorders.

Scientists at Cognigenics have pioneered RNA-based interventions with research aimed at modulating gene expression using RNA interference (RNAi) and antisense oligonucleotides. This work is anchored in addressing critical challenges in neuropsychiatry, such as specificity and durability of therapeutic effects, often lacking in traditional pharmacological approaches.

Targeting Anxiety and Cognitive Disorders

One of Cognigenics' breakthroughs lies in its targeting of the HTR2a gene, which encodes the 5-HT2a receptor, a key player in modulating neural excitability and behavior. Anxiety spectrum disorders, characterized by heightened fear and stress responses, have been a cornerstone of the company's research. By employing RNA-based methods, they developed a platform to downregulate receptor expression, demonstrating reductions in anxiety-like behaviors in animal models.

Dr. Tacy Brandmeyer has Scientists are thinking about cognitive disorders, particularly mild cognitive impairment and early-onset Alzheimer's, Cognigenics' extends to modulating neuronal plasticity and amyloid-beta pathways. These interventions, targeting root molecular dysfunctions, aim to alter disease progression—a departure from traditional therapies that manage symptoms not underlying mechanisms.

Non-invasive Gene Delivery

Cognigenics has successfully utilized adeno-associated virus serotype 9 (AAV9) vectors delivered intranasally. This innovative approach bypasses the blood-brain barrier, a long-standing challenge in CNS therapeutics, and enables localized genetic modulation with high precision.

Behavioral and Molecular Outcomes

Experimental results underscore significant reductions in anxiety-like behaviors, enhancements in memory performance, and decreased excitability in primary cortical neurons. These findings highlight the dual benefits of RNA therapeutics in both functional restoration and behavioral improvement.

Specificity and Safety

By targeting specific genes like HTR2a, Cognigenics' platform minimizes off-target effects, a key limitation of traditional systemic drugs. Preclinical data demonstrate robust safety profiles and durable therapeutic effects.

Implications for the Future of Neuropsychiatry

The insights gained from these studies go beyond isolated successes—they lay the foundation for a paradigm shift in treatment. RNA-based therapeutics, promoting durable outcomes, with long-lasting effects that reduce the need for frequent dosing and enhance patient compliance; and facilitating disease modulation, addressing the underlying mechanisms of neuropsychiatric disorders.

Challenges and Considerations

While these advances are promising, several hurdles remain for clinical translation. Key challenges include ensuring large-scale manufacturing of RNA-based therapies, addressing regulatory pathways, and optimizing delivery systems for diverse patient populations. Additionally, the integration of RNA therapeutics into existing treatment paradigms will require deep collaboration between researchers, clinicians, and policymakers.

Conclusion

The work presented here places Cognigenics at the forefront of RNA-based therapeutics for mental health. These studies collectively represent a transformative approach to understanding and treating complex neuropsychiatric disorders at their molecular roots. As the field progresses, the continued exploration of RNA therapies holds the promise of reshaping mental health care, offering hope for more precise and effective treatments for some of the most challenging conditions in medicine.

27. Paradigm Shift

ARTICLE TITLE

Intranasal Delivery of shRNA to Knockdown the 5-HT2a Receptor Enhances Memory and Alleviates Anxiety.

Authors: Troy T. Rohn[1,2], Dean Radin[2], Tracy Bran meyer[2], Peter G. Seidler[2], Barry J. Linder[2], Tom Lytle[2], John L. Mee[2] and Fabio Macciardi[2,3]

Abstract: Short-hairpin RNAs (shRNA), targeting knockdown of specific genes, hold enormous promise for precision-based therapeutics to treat numerous neurodegenerative disorders. However, whether shRNA constructed molecules can modify neuronal circuits underlying certain behaviors has not been explored. We designed shRNA to knockdown the human HTR2A gene in vitro using iPSCdifferentiated neurons. Multi-electrode array (MEA) results showed that the knockdown of the 5-HT2a mRNA and receptor protein led to a decrease in spontaneous electrical activity. In vivo, intranasal delivery of AAV9 vectors containing shRNA resulted in a decrease in anxiety-like behavior in mice and a significant improvement in memory in both mice (104%) and rats (92%) compared to vehicle-treated animals. Our demonstration of a non-invasive shRNA delivery platform that can bypass the blood–brain barrier has broad implications for treating numerous neurological mental disorders. Specifically, targeting the HTR2A gene presents a novel therapeutic approach for treating chronic anxiety and age-related cognitive decline.

Translational Psychiatry (2024) 14:154;
https://doi.org/10.1038/s41398-024-02879-y

The following page is the first page of that study. The complete study is available at the above link.

Translational Psychiatry www.nature.com/tp

ARTICLE OPEN

Check for updates

Intranasal delivery of shRNA to knockdown the 5HT-2A receptor enhances memory and alleviates anxiety

Troy T. Rohn [1,2]✉, Dean Radin[2], Tracy Brandmeyer[2], Peter G. Seidler[2], Barry J. Linder[2], Tom Lytle[2], John L. Mee[2] and Fabio Macciardi[2,3]

Short-hairpin RNAs (shRNA), targeting knockdown of specific genes, hold enormous promise for precision-based therapeutics to treat numerous neurodegenerative disorders. However, whether shRNA constructed molecules can modify neuronal circuits underlying certain behaviors has not been explored. We designed shRNA to knockdown the human *HTR2A* gene in vitro using iPSC-differentiated neurons. Multi-electrode array (MEA) results showed that the knockdown of the 5HT-2A mRNA and receptor protein led to a decrease in spontaneous electrical activity. In vivo, intranasal delivery of AAV9 vectors containing shRNA resulted in a decrease in anxiety-like behavior in mice and a significant improvement in memory in both mice (104%) and rats (92%) compared to vehicle-treated animals. Our demonstration of a non-invasive shRNA delivery platform that can bypass the blood–brain barrier has broad implications for treating numerous neurological mental disorders. Specifically, targeting the *HTR2A* gene presents a novel therapeutic approach for treating chronic anxiety and age-related cognitive decline.

Translational Psychiatry (2024)14:154; https://doi.org/10.1038/s41398-024-02879-y

INTRODUCTION

Neurological disorders such as Alzheimer's disease (AD) and chronic anxiety are a major public mental health challenge, affecting millions of people worldwide. However, despite significant research efforts, there has been limited success in treating the symptoms associated with these disorders. Precision-based therapeutics such as CRISPR/Cas9 and RNA interference molecules offer a promising new approach to treating neurological and neurodegenerative disorders. shRNA represents one class of RNA interference molecules that has a mechanism based on the sequence-specific degradation of host mRNA through cytoplasmic delivery and degradation of double-stranded RNA through the RISC pathway [1, 2]. Whereas CRISPR/Cas9 leads to permanent changes in the genome, shRNA induces reversible gene silencing through a posttranslational regulatory process targeting degradation of specific mRNAs. Besides being reversible, shRNA has also been widely used in research for over a decade and there are currently four FDA-approved therapeutics that use shRNA to treat rare metabolic disorders [3]. Moreover, shRNA targets a specific mRNA sequence, meaning it can potentially distinguish between closely related genes with high sequence homology. However, whether shRNA could be used to modify certain behavioral traits within the CNS has not been investigated.

Currently there is a great need for non-invasive methods of delivering gene therapy to the brain. shRNA represents a potential powerful tool, but it is difficult to deliver to the brain because of the blood–brain barrier (BBB). The BBB is a protective layer that prevents most molecules from entering the brain [4]. One way to overcome this challenge is to use adeno-associated viral (AAV) vectors. AAV vectors are viruses that have been modified to be safe and effective for gene delivery due to their ability to deliver stable, long-lasting transgene expression in non-dividing cells [5].

We recently demonstrated that intranasal delivery of CRISPR/Cas9 encapsulated within adeno-associated viral (serotype AAV9) vectors could bypass the BBB and lead to knockout of the *HTR2A* gene in neuronal populations [6] (US Patent Application No. 63/283,150). The *HTR2A* gene encodes for the 5HT-2A receptor, one of the fifteen serotonin receptor subtypes expressed in the brain, and is implicated in both anxiety disorders [7, 8] and memory [9–11]. In the present study, we designed a shRNA to knockdown the *HTR2A* gene and demonstrate a decrease in spontaneous electrical activity in human iPSC-differentiated neurons in vitro as well as enhanced memory and a reduction in anxiety in mice and rats in vivo. The development of this non-invasive shRNA delivery platform, which is capable of bypassing the blood–brain barrier, holds substantial implications for treatment of a wide spectrum of neurological and neurodegenerative disorders. Specifically, the targeting of the *HTR2A* gene emerges as a novel and promising therapeutic approach for addressing conditions such as chronic anxiety, mild cognitive impairment, dementia, and possibly AD.

MATERIALS AND METHODS

Guide RNA and AAV9 vector design for CRISPR/Cas9 experiments

Details on the methods used to synthesize, design, and validate guide RNA (gRNA) and knockdown of the mouse *HTR2A* gene have been previously reported [6]. In brief, the selected gRNA, TGCAATTAGGTGACGACTCGAGG (US Patent Application No. 63/283,150), would give no predicted off-target cut sites, produce an 86.6% frameshift frequency, and a precision score of 0.55. Two different adeno-associated virus serotype 9 (AAV9) vectors were

28. Blueprint for Tomorrow's Medicine

ARTICLE TITLE

Treatment with shRNA to knockdown the 5-HT2a receptor improves memory in vivo and decreases excitability in primary cortical neurons.

Authors: Troy T. Rohn[1,2], Dean Radin[2], Tracy Brandmeyer[2], Peter G. Seidler[2], Barry J. Linder[2], Tom Lytle[2], John L. Mee[2] and Fabio Macciardi[2,3]

Abstract: This paper outlines the potential of RNA-based therapeutics to deliver long-term benefits, providing an alternative to current treatment options like SSRIs that are often associated with side effects and limited efficacy. This third article sets a vision for how these RNA therapies can reshape not only individual patient outcomes but also the broader field of neuropsychiatry by offering treatments that are both specific and durable.

This forward-looking perspective sparks the imagination, suggesting we are on the cusp of a new era in which mental health disorders are addressed at their most fundamental molecular levels, allowing for more sustainable and profound changes in patient well-being.

Genomic Psychiatry (2024), https://doi.org/10.61373/gp024r.0043

The following page is the first page of the paper.

gp.genomicpress.com

Genomic Psychiatry

OPEN

RESEARCH REPORT

Treatment with shRNA to knockdown the 5-HT2A receptor improves memory in vivo and decreases excitability in primary cortical neurons

Troy T. Rohn[1], Dean Radin[1], Tracy Brandmeyer[1], Peter G. Seidler[1], Barry J. Linder[1], Tom Lytle[1], David Pyrce[1], John L. Mee[1], and Fabio Macciardi[1,2]

[1]Cognigenics, Eagle, Idaho, 83616, USA
[2]Department of Psychiatry and Human Behavior, University of California, Irvine, California 92697, USA

Corresponding Author: Troy T. Rohn, 1372 S. Eagle Road, Suite 197 Eagle, Idaho 83616 USA. E-mail: troy.rohn@cognigenics.io

Genomic Psychiatry; https://doi.org/10.61373/gp024r.0043

Short hairpin RNAs (shRNA), targeting knockdown of specific genes, hold enormous promise for precision-based therapeutics to treat numerous neurodegenerative disorders. We designed an AAV9-shRNA targeting the downregulation of the 5-HT2A receptor, and recently demonstrated that intranasal delivery of this shRNA (referred to as COG-201), decreased anxiety and enhanced memory in mice and rats. In the current study, we provide additional in vivo data supporting a role of COG-201 in enhancing memory and functional in vitro data, whereby knockdown of the 5-HT2A receptor in primary mouse cortical neurons led to a significant decrease in mRNA expression ($p = 0.0007$), protein expression p-value = 0.0002, and in spontaneous electrical activity as measured by multielectrode array. In this regard, we observed a significant decrease in the number of spikes (p-value = 0.002), the mean firing rate (p-value = 0.002), the number of bursts (p-value = 0.015), and a decrease in the synchrony index (p-value = 0.005). The decrease in mRNA and protein expression, along with reduced spontaneous electrical activity in primary mouse cortical neurons, corroborate our in vivo findings and underscore the efficacy of COG-201 in decreasing *HTR2A* gene expression. This convergence of in vitro and in vivo evidence solidifies the potential of COG-201 as a targeted therapeutic strategy. The ability of COG-201 to decrease anxiety and enhance memory in animal models suggests that similar benefits might be achievable in humans. This could lead to the development of new treatments for conditions like generalized anxiety disorder, post-traumatic stress disorder (PTSD), and cognitive impairments associated with aging or neurodegenerative diseases.

Keywords: RNA interference, 5-HT2A receptor, memory enhancement, neuronal excitability, anxiety, cognitive impairment.

Introduction

Neurological disorders such as mild cognitive impairment (MCI) and chronic anxiety are a major public mental health challenge, affecting millions of people worldwide. MCI is often a transitional stage between healthy aging and dementia. Depending on the inclusion criteria, the prevalence of MCI has been estimated to be between 5.0% and 36.7% (1). According to a systematic review and meta-analysis, the overall pooled prevalence of anxiety in patients with MCI is approximately 21%. This prevalence rate varies based on the source of the sample and the method of diagnosis. For example, the prevalence of anxiety in community-based samples of patients with MCI is about 14.3%, while it is approximately 31.2% in clinic-based samples (2). Based on these statistics, we estimate that roughly 1.5–2 million Americans suffer from MCI with an underlying anxiety disorder. Currently, there is no single medication to treat both cognitive impairments and anxiety in this patient population.

Precision-based therapeutics such as RNA interference offer a promising new approach to treating neurological and neurodegenerative disorders. Short hairpin RNA (shRNA) represents one class of RNA interference molecules that has a mechanism based on the sequence-specific degradation of host mRNA through cytoplasmic delivery and degradation of double-stranded RNA through the RNA-induced silencing complex (RISC) pathway (3, 4). We designed plasmids containing the RNA instructions to construct a specific shRNA to silence the *HTR2A* gene (U.S. Patent Application No. 63/567,853). The *HTR2A* gene encodes for the 5-HT2A receptor, one of the 15 known serotonin receptor subtypes expressed in the brain, and is implicated in both anxiety disorders (5, 6) and memory (7–9). This plasmid contains a neuronal specific promoter, MeCP2 and is packaged within AAV9 viral particles. Intranasal treatment of this AVV9-shRNA (herein termed COG-201) in mice or rats significantly decreased anxiety and improved memory (10). In this study, we present further evidence supporting the memory-enhancing effects COG-201. We also provide functional data from experiments on primary cortical neurons taken from mice. Our results show that treatment with COG-201 leads to reduced spontaneous electrical activity in these neurons. This effect occurs specifically after reducing the expression of the 5-HT2A receptor. These findings bolster the potential of intranasal shRNA delivery as a non-invasive therapeutic method and establish a foundation for continued investigation into its role in treating anxiety and cognitive deficits linked to a spectrum of neurodegenerative diseases.

Methods

shRNA Design and AAV9 Vector Design

Construction of the mouse shRNA to target knockdown of the 5-HT2A receptor was as previously described (11). The mouse *HTR2A* gene consists of three exons that give rise to two major isoforms and is found on chromosome 13. The predicted binding region of the primary RNA transcript for this sequence is the beginning of exon 2, which would lead to the potential knockdown of all possible isoforms. The following sequence was used for assembly of the shRNA based on *in vitro* testing indicating a 77% knockdown:

GCTGAGCACATCCAGGTAAATCCAGGTTTTGGCCACGACTGACCTGGATTT CTGGATGTGCT CAG

No knockdown was observed with the empty vector control or a scrambled shRNA control (Figure 2B). For validation and screening, knockdown was verified using HEK293 cells cotransfected with the cDNA plasmid containing the *HTR2A* gene target. For in vitro treatment of primary mouse cortical neurons, shRNA delivery subcloning of the shRNA was carried out in a modified pAAV *cis*-plasmid under the neuronal-specific promoter, MeCP2. The inclusion of the MeCP2 promoter is a crucial element design, as it ensures expression of the shRNA plasmid only in neuronal populations. A reporter gene enhanced green fluorescent protein was subcloned upstream of the shRNA sequence. AAV9 viral large-scale transfection of plasmids was carried out in HEK293 cells and purified through a series of CsCl centrifugations. Titer load (in genome copy number per mL, or GC/mL) was determined through quantitative real-time PCR, with typical yields giving 1–2 × 10^{13} GC/mL. All AAV9 vectors were stored in phosphate buffered saline (PBS) with 5% glycerol at −80°C until used. Design, manufacturing, and purification of AAV9 vectors used in this study were performed by Vector Biolabs (Malvern, PA).

Novel Object Recognition Test

The object recognition task is used to assess short-term memory, intermediate-term memory, and long-term memory in rats and was performed as previously described (11). The task is based on the natural tendency of rats to preferentially explore a novel versus a familiar object, which requires memory of the familiar object. The time delay design allows for the screening of compounds with potential cognitive enhancing

29. The Dawn of Precision Medicine for Mental Health

ARTICLE TITLE

Genetic Modulation of the HTR2A Gene Reduces Anxiety-Related Behavior in Mice.

Authors: Troy T. Rohn[a,b], Dean Radin[b], Tracy Brandmeyer[b], Barry J. Linder[b], Emile Andriambeloson[c], Stéphanie Wagner[c], James Kehler[d], Ana Vasileva[d], Huaien Wang[d], John L. Mee[b] and James H. Fallon[b,e]

Abstract: The expanding field of precision gene editing using CRISPR/Cas9 has demonstrated its potential as a transformative technology in the treatment of various diseases. However, whether this genome-editing tool could be used to modify neural circuits in the central nervous system (CNS), which are implicated in complex behavioral traits, remains uncertain. In this study, we demonstrate the feasibility of noninvasive, intranasal delivery of adeno-associated virus serotype 9 (AAV9) vectors containing CRISPR/Cas9 cargo within the CNS resulting in modification of the HTR2A receptor gene. In vitro, exposure to primary mouse cortical neurons to AAV9 vectors targeting the HT2RA gene led to a concentration-dependent decrease in spontaneous electrical activity following multielectrode array (MEA) analysis. In vivo, at 5 weeks postintranasal delivery in mice, analysis of brain samples revealed single base pair deletions and nonsense mutations, leading to an 8.46-fold reduction in mRNA expression and a corresponding 68% decrease in the 5HT-2A receptor staining. Our findings also demonstrate a significant decrease in anxiety-like behavior in treated mice. This study constitutes the first successful demonstration of a noninvasive CRISPR/Cas9 delivery platform, capable of bypassing the blood–brain barrier and enabling modulation of neuronal 5HT-2A receptor pathways. The results of this study targeting the HTR2A gene provide a foundation for the development of innovative therapeutic strategies for a broad range of neurological disorders, including anxiety, depression, attentional deficits, and cognitive dysfunction.

PNAS Nexus (2023) 2:6;
https://doi.org/10.1093/pnasnexus/pgad170

The following page is the first page of the study.

PNAS nexus

PNAS Nexus, 2023, 2, 1–13
https://doi.org/10.1093/pnasnexus/pgad170
Advance access publication 20 June 2023
Research Report

Genetic modulation of the HTR2A gene reduces anxiety-related behavior in mice

Troy T. Rohn[a,b,*], Dean Radin[b], Tracy Brandmeyer[b], Barry J. Linder[b], Emile Andriambeloson[c], Stéphanie Wagner[c], James Kehler[d], Ana Vasileva[d], Huaien Wang[d], John L. Mee[b] and James H. Fallon[b,e]

[a]Department of Biological Sciences, Boise State University, Boise, ID 83725, USA
[b]Cognigenics, Eagle, ID 83616, USA
[c]Neurofit, Illkirch-Graffenstaden 67400, France
[d]Mirimus Inc., Brooklynn, NY 11226, USA
[e]Department of Psychiatry and Human Behavior, University of California, Irvine, CA 92697, USA
*To whom correspondence should be addressed: Email: trohn@boisestate.edu
Edited By: Andrey Abramov

Abstract

The expanding field of precision gene editing using CRISPR/Cas9 has demonstrated its potential as a transformative technology in the treatment of various diseases. However, whether this genome-editing tool could be used to modify neural circuits in the central nervous system (CNS), which are implicated in complex behavioral traits, remains uncertain. In this study, we demonstrate the feasibility of noninvasive, intranasal delivery of adeno-associated virus serotype 9 (AAV9) vectors containing CRISPR/Cas9 cargo within the CNS resulting in modification of the *HTR2A* receptor gene. In vitro, exposure to primary mouse cortical neurons to AAV9 vectors targeting the *HT2RA* gene led to a concentration-dependent decrease in spontaneous electrical activity following multielectrode array (MEA) analysis. In vivo, at 5 weeks postintranasal delivery in mice, analysis of brain samples revealed single base pair deletions and nonsense mutations, leading to an 8.46-fold reduction in mRNA expression and a corresponding 68% decrease in the 5HT-2A receptor staining. Our findings also demonstrate a significant decrease in anxiety-like behavior in treated mice. This study constitutes the first successful demonstration of a noninvasive CRISPR/Cas9 delivery platform, capable of bypassing the blood–brain barrier and enabling modulation of neuronal 5HT-2A receptor pathways. The results of this study targeting the *HTR2A* gene provide a foundation for the development of innovative therapeutic strategies for a broad range of neurological disorders, including anxiety, depression, attentional deficits, and cognitive dysfunction.

Keywords: CRISPR, HTR2A, 5HT-2A receptor, anxiety, intranasal delivery

Significance Statement

Current therapies for anxiety and depression rely heavily on selective serotonin reuptake inhibitors that may impact numerous serotonergic receptor pathways, are fraught with side effects, and require daily dosing. Herein, we employed precision gene targeting to selectively knockdown the 5HT-2A receptor, one of the 15 serotonin receptor subtypes known to play a key role in anxiety and depression. To accomplish this goal, we used a noninvasive, intranasal delivery platform consisting of two adeno-associated virus (AAV) vectors containing plasmids for CRISPR/Cas9 and guide RNA, respectively, to knockdown the 5HT-2A receptors in anatomical and functional connectomes implicated in both anxiety and depression. Our results indicated administered AAV–CRISPR–Cas9 significantly reduced anxiety, thus demonstrating that complex behavioral traits can be directly modified long-term through genetic editing.

Introduction

CRISPR is a powerful tool that has the potential to revolutionize genomic engineering by potentially treating various diseases (1). The CRISPR/Cas9 system consists of two parts: the Cas9 protein which is the nuclease capable of inducing double-strand DNA breaks and the guide RNA (gRNA) which is designed to target any specific DNA sequence in the genome. Once double-strand breaks are generated, the DNA is repaired by either nonhomologous end joining (NHEJ) or homologous-directed repair (HDR) pathways that induce insertions or deletions (indels) at the target site (2). Dozens of human clinical trials are currently underway using the CRISPR/Cas system, including those for sickle cell

OXFORD UNIVERSITY PRESS

Competing Interest: J.L.M. and D.R. are cofounders of Cognigenics and members of its scientific advisory board and hold equity in the company. T.T.R. is a part-time consultant serving as Director of Preclinical Research at Cognigenics and, in addition to receiving a salary, owns shares of the company's common stock and options for common shares. J.H.F. is a part-time consultant serving as Chief Science Officer at Cognigenics, Inc., and is a member of its scientific advisory board. In addition to receiving a salary, he owns shares of the company's common stock. He is the inventor of a patent application entitled "Systems and Methods for an Intranasal Drug Delivery System." All other authors declare no competing interests.
Received: March 4, 2023. **Accepted:** May 15, 2023

30. Executive Summary

Cognitive Genetics

Innovating Mental Health and Cognitive Optimization

Cognitive Genetics explores the emerging field of genetic neuroengineering and its potential to transform mental health treatments and cognitive enhancement. The book, written by Peter Seidler with contributions from Dean Radin and J. L. Mee, covers advances in RNA-based therapeutics, particularly their applications in treating neuropsychiatric and neurodegenerative disorders.

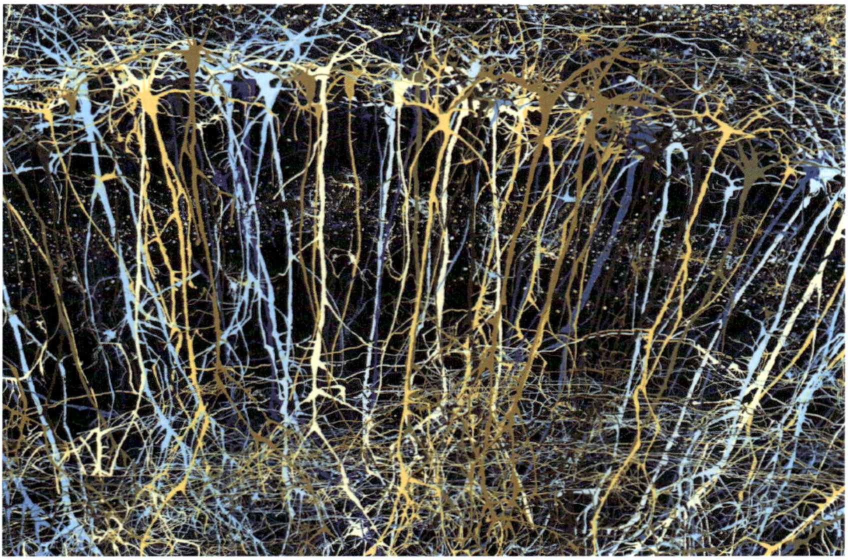

BBHC-Blue-and-Gold, Greg Dunn

The text offers a forward-looking perspective on how precision gene-editing techniques like RNA interference (RNAi) are poised to revolutionize mental health care by addressing neuropsychiatric and neurocognitive conditions at the molecular level.

The Promise of RNA Therapeutics

The book emphasizes the groundbreaking potential of RNA-based therapies, particularly short hairpin RNA (shRNA) and RNA interference technologies, which can selectively silence gene expression.

These approaches are presented as future alternatives to conventional psychotropic medications like SSRIs, with significant side effects and limited efficacy. RNA therapeutics, by contrast, offer the possibility of precisely targeting dysfunctional neural circuits implicated in conditions such as anxiety, depression, mild cognitive impairment (MCI), and even Alzheimer's disease.

Precision and Noninvasive Delivery

RNA-based therapies offer high precision targeting specific neuronal receptors like 5-HT2a, crucial in anxiety and memory regulation, with minimal off-target effects. Noninvasive delivery, such as intranasal administration, bypasses the blood-brain barrier, making CNS treatment safer and more efficient than invasive methods.

Applications for Neuropsychiatric Conditions

Cognitive Genetics delves into the potential applications of RNA therapeutics for various conditions. In addition to treating anxiety and cognitive impairment, the text explores the possibilities for addressing more complex neurodegenerative conditions like Alzheimer's, schizophrenia, and epilepsy. The research underscores how RNA-based therapies could slow or reverse the progression of these diseases, offering hope for more effective and long-lasting treatments.

Ethical Considerations and Societal Impact

The book also considers the profound ethical implications of genetic interventions in mental health. While the therapeutic potential is immense, Seidler and his contributors caution against unintended consequences, such as creating "superhumans" or misusing gene-editing technologies in areas like military applications. They advocate for a balanced approach, emphasizing the need for ethical oversight and collaboration across scientific disciplines, as well as with policymakers and patient advocacy groups.

Looking Ahead

Seidler, Radin, and Mee predict that RNA-based therapies will provide better mental health treatments and enhance cognitive functioning in healthy individuals. Genetic interventions' potential to boost memory, learning, and overall mental capacity represents a new frontier in neuroscience and genetic medicine. As these therapies progress through clinical trials, *Cognitive Genetics* anticipates a future where gene-editing and RNA-based treatments are integrated into mainstream medical practice, reshaping mental health care and human cognitive potential.

i. Notes to the Introduction

By J. L. Mee

Marketing Strategy

Market size and attributes: The long-range U.S. cognitive enhancement market is primarily composed of Generation Z. This generation's share of the U.S. population is projected to reach 25% in 2025, while in the EU, it is expected to rise to 21%.

Generation Z market: Generation Z faces unprecedented challenges in career development as automation and artificial intelligence replace conventional jobs with new, more cognitively demanding roles. As we enter the Fourth Industrial Revolution, artificial intelligence and robotics will eliminate millions of jobs in the next decade.

Market psychographics: Genetic cognitive upgrades will become popular in the 2020s in a world where cannabis is widely legalized and cognitively expanded states are commonplace

Generation	Need
Generation Z	Expanded human value skills such as imagination and creativity to stay ahead of competition for jobs from AI and automation.
Generation Z Parents	Give their children every advantage in tomorrow's challenging economy.
Entrepreneurs	More innovation, creativity, energy, and focus to overcome global competition.
Personal Growth Seekers	Higher consciousness, greater clarity of purpose, and accelerated personal evolution.
Spouses	Greater harmony in relationships, love, and emotional intelligence.
Knowledge Workers	More imagination, intuition, and mental acuity to boost performance.
Families	Increased empathy, mindfulness, love, compassion, and harmony.
Adult Learners	Develop new skills faster and improve academic performance.

Summary of Generation Z's & ancillary markets' need. Source: J. L. Mee

Social Responsibility

Cognitively enhanced individuals will make ideal employees for companies competing in the 21st Century knowledge economy, due to their enriched cognitive capital for developing creative solutions to business problems, and higher emotional balance and morale. Hence, the upgrades will have an economic value to employers.

Candidates for upgrades make an investment in themselves which pays dividends throughout their careers in the form of higher salaries and faster advancement. For professionals, these rewards could be comparable to the value of a college degree, and for entrepreneurs, potentially greater.

Philanthropy: The socially responsible implementation of cognitive engineering requires special measures to level the playing field to avoid exacerbating existing inequalities or creating a cognitive elite. Accordingly, a non-profit organization will manage a cognitive enhancement scholarship fund for deserving, underprivileged individuals of high merit. The influence of cognitively enhanced individuals will spread through their families, communities and workplaces, ultimately benefiting humanity as a whole.

Families: Expanded consciousness fosters harmony in human relationships. It enriches the spiritual dimension of families, bringing more awareness to family dynamics, and creating deeper and more meaningful relationships between spouses.

Communities: Greater peace of mind and emotional balance will reduce conflicts individually and societally and improve community coherence.

Workplaces: Employees with enriched cognitive capital can help companies develop creative solutions to business problems. Increased happiness and positive emotional states will also engender greater employee wellness by upregulating epigenetic pathways.

Humanity: A global shift towards higher consciousness will help people to realize their interconnectedness with the human community and nature. Heightened cognitive abilities can be applied to solving some of the planet's biggest challenges and problems. Greater awareness of humankind's spiritual dimension can open the way to a golden age of wisdom and foster accelerated progress in the arts and sciences.

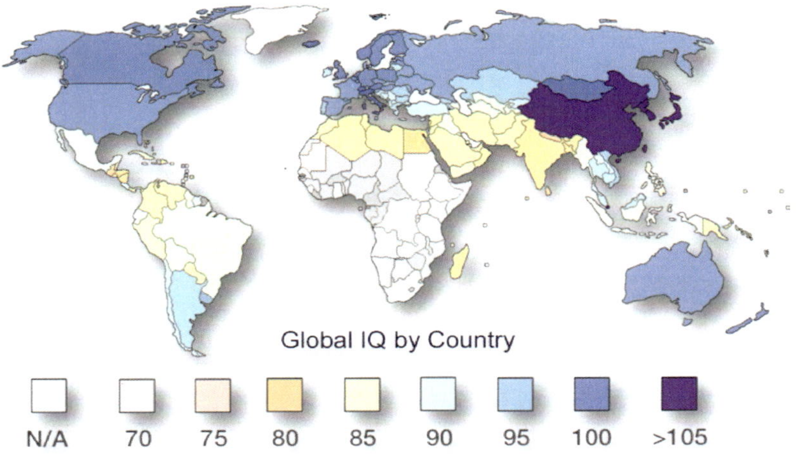

IQ and Global Inequality by Richard Lynn and Tatu Vanhanen

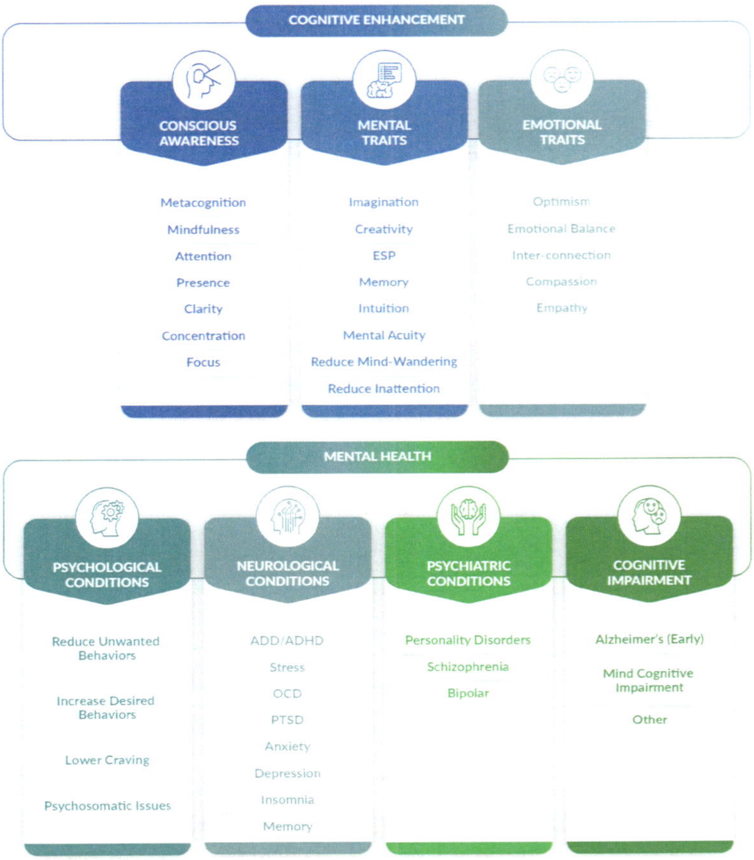

Source: J. L. Mee

ii. Selected Sources, by Chapter

Preface

Harari, Yuval. *Homo Deus: A Brief History of Tomorrow.* 2015.

Rohn, Troy T., et al.. "Genetic modulation of the HTR2A gene reduces anxiety-related behavior in mice." *Journal of Alzheimer's disease & Parkinsonism,* 2023.

McCarthy, M. M., et al.. "An amino-terminal fragment of apolipoprotein E4 leads to behavioral deficits, increased PHF-1 immunoreactivity, and mortality in zebrafish." *Journal of Alzheimer's disease & Parkinsonism,* 2022

Foreword

Plomin, R., & von Stumm, S. (2018). The new genetics of intelligence. *Nature Reviews Genetics*, 19(3), 148-159.

Kendler, K. S., & Prescott, C. A. (2006). *Genes, environment, and psychopathology: Understanding the causes of psychiatric and substance use disorders.* New York: Guilford Press.

Nuffield Council on Bioethics. (2018). *Genome editing and human reproduction: Social and ethical issues.* London: Nuffield Council on Bioethics.

Harris, J. (2011). *Enhancing evolution: The ethical case for making better people.* Princeton University Press.

Tost, H., Champagne, F. A., & Meyer-Lindenberg, A. (2015). Environmental influence in the brain, human welfare and mental health. *Nature Neuroscience*, 18(10), 1421-1431.

Deary, I. J., Penke, L., & Johnson, W. (2010). The neuroscience of human intelligence differences. *Nature Reviews Neuroscience,* 11(3), 201-211.

Savulescu, J., & Bostrom, N. (Eds.). (2009). *Human enhancement.* Oxford University Press.

Introduction

Al-Rodhan, Nayef. Brain Gain: The Emerging Security and Ethical Challenges of Cognitive Enhancement. *Basingstoke*: Palgrave Macmillan, 2015.

Brandmeyer, Tracy, and Arnaud Delorme. "Closed-Loop Frontal Midline Neurofeedback: A Novel Approach for Training Focused-Attention Meditation." *Journal of Neurotherapy,* vol. 24, no. 2, 2020, pp. 100-112.

CoreSight Research. "Gen Z: Get Ready for the Most Self-Conscious, Demanding Consumer Segment." *CoreSight Research,* 2017.

Fortune.com. "Digital Health Care Revolution." Fortune, 12 Jan. 2017, www.fortune.com/digital-health-care-revolution.

Gizmodo. "China Has Already Edited 86 People With CRISPR." *Gizmodo*, 28 June 2018, www.gizmodo.com/china-edited-86-people-with-crispr.

Lynn, Richard, Tatu Vanhanen, and M. Stuart. IQ and Global Inequality. Augusta, GA: *Washington Summit Publishers,* 2006.

McKinsey Global Institute. A Future That Works: Automation, Employment and Productivity. *McKinsey & Company,* 2017.

Mellon, Jim, Al Chalabi, and Andrew Craig. Cracking the Code: Understand and Profit from the Biotech Revolution That Will Transform Our Lives and Generate Fortunes. London: *Wiley*, 2012.

Reggente, Nicolas, et al. "Decoding Depth of Meditation: EEG Insights from Expert Vipassana Practitioners." *NeuroImage*, vol. 242, 2024, p. 118418.

Rohn, Troy T., et al. "Long-Lasting Genetic Therapeutics for Debilitating Neurological Conditions." *Nature Neuroscience,* vol. 27, 2024, pp. 121-133.

WIRED. "Easy DNA Editing Will Remake the World. Buckle Up." *WIRED*, 11 July 2015, www.wired.com/easy-dna-editing-will-remake-the-world.

Part One: Social

1. A Quest for Innovative Treatments

Gozzo, L., Spina, E., & Drago, F. (2023). Innovative treatments for neuro-psychiatric diseases. *Frontiers in Neuroscience,* 17. https://doi.org/10.3389/fnins.2023.1247681

Maroney, N. J. (2023). Strategies to address challenges in neuroscience drug discovery and development. *International Journal of Neuropsychopharmacology,* 26(1), 18-29. https://doi.org/10.1093/ijnp/pyad005

Smith, J. A., & Nguyen, T. (2023). Advancements in clinical RNA therapeutics: Present developments and future prospects. *Trends in Pharmacological Sciences,* 44(7), 502-519. https://doi.org/10.1016/j.tips.2023.04.007

Zogg, H., Singh, R., & Ro, S. (2022). Current advances in RNA therapeutics for human diseases. *International Journal of Molecular Sciences,* 23(5), 2736. https://doi.org/10.3390/ijms23052736

2. A New Class of Genetic Therapeutics for Mental Health By Dean Radin Ph.D.

Anagha, K., Shihabudheen, P., & Uvais, N. A. (2021). Side effect profiles of selective serotonin reuptake inhibitors: A cross-sectional study in a naturalistic setting. *Primary Care Companion for CNS Disorders,* 23(4), 20m02747. doi: 10.4088/PCC.20m02747. PMID: 34324797.

Anderson, N. D. (2019). State of the science on mild cognitive impairment (MCI). *CNS Spectrums,* 24(1), 78-87. doi: 10.1017/S1092852918001347. Epub 2019 Jan 17. PMID: 30651152.

Garcia-Romeu, A. et al.., (2022). Psychedelics as novel therapeutics in Alzheimer's disease: Rationale and potential mechanisms. *Current Topics in Behavioral Neurosciences*, 56, 287-317. doi:10.1007/7854_2021_267. PMID: 34734390.

Hakimi, Y., et al.., (2020). Paradoxical adverse drug reactions: Descriptive analysis of French reports. *European Journal of Clinical Pharmacology,* 76(8), 1169-1174. doi: 10.1007/s00228-020-02892-2. Epub 2020 May 16. PMID: 32418024.

Lewis, V., et al.., (2023). A non-hallucinogenic LSD analog with therapeutic potential for mood disorders. *Cell Reports,* 42(3), 112203. doi: 10.1016/j.celrep.2023.112203. Epub 2023 Mar 6. PMID: 36884348; PMCID: PMC10112881.

Orgeta, V., et al.., (2022). Psychological treatments for depression and anxiety in dementia and mild cognitive impairment. *Cochrane Database of Systematic Reviews,* 4(4), CD009125. doi: 10.1002/14651858.CD009125.pub3. PMID: 35466396; PMCID: PMC9035877.

Raj, P., Rauniyar, S., & Sapkale, B. (2023). Psychedelic drugs or hallucinogens: Exploring their medicinal potential. *Cureus*, 15(11), e48719. doi: 10.7759/cureus.48719. PMID: 38094517; PMCID: PMC10716812.

Rohn, T. et al.., (2023). Genetic modulation of the HTR2A gene reduces anxiety-related behavior in mice. *PNAS Nexus,* 2(6), pgad170. doi: 10.1093/pnasnexus/pgad170. PMID: 37346271; PMCID: PMC10281383.

van den Berg, M. et al.., (2022). How to account for hallucinations in the interpretation of the antidepressant effects of psychedelics: A translational framework. *Psychopharmacology* (Berl), 239(6), 1853-1879. doi: 10.1007/s00213-022-06106-8. Epub 2022 Mar 29. PMID: 35348806; PMCID: PMC9166823.

3. Precision & Promise: RNA Neuromodulation

Brandmeyer, T., & Delorme, A. (2021). Meditation and the wandering mind: A theoretical framework of underlying neurocognitive mechanisms. *Neuroscience & Biobehavioral Reviews,* 131, 803-812. https://doi.org/10.1016/j.neubiorev.2021.10.010

Cannard, C., Brandmeyer, T., Wahbeh, H., & Delorme, A. (2020). Self-health monitoring and wearable neurotechnologies. *Frontiers in Human Neuroscience,* 14, 214. https://doi.org/10.3389/fnhum.2020.00214

Gozzo, L., Spina, E., & Drago, F. (2023). Innovative treatments for neuro-psychiatric diseases. *Frontiers in Neuroscience*, 17. https://doi.org/10.3389/fnins.2023.1247681

Maroney, N. J. (2023). Strategies to address challenges in neuroscience drug discovery and development. *International Journal of Neuropsychopharmacology*, 26(1), 18-29. https://doi.org/10.1093/ijnp/pyad005

4. Safety, Efficacy, and Patient Outcomes

Coonse, K. G., Dickson, D. W., & Hardy, J. A. (2012). Amino-terminal cleavage of apolipoprotein E4 in Alzheimer's disease. *Neurobiology of Aging,* 33(10), 2345-2357. https://doi.org/10.1016/j.neurobiolaging.2012.05.003

Delorme, A., & Brandmeyer, T. (2019). When the meditating mind wanders: Influence of meditation depth, continuity, and practice type on the occurrence of mind-wandering. *Consciousness and Cognition,* 73, 102746. https://doi.org/10.1016/j.concog.2019.03.015

Rohn, T. T., *Radin*, D., Brandmeyer, T., Seidler, P. G., Linder, B. J., Lytle, T., Mee, J. L., & Macciardi, F. (2024). Intranasal delivery of shRNA to knockdown the 5-HT2a receptor enhances memory and alleviates anxiety. Translational Psychiatry, 14(154). https://doi.org/10.1038/s41398-024-02135-x

5. Mechanics to Genetics: Remembering Technological Epochs

Cribbs, A. P., & Perera, S. M. (2017). Ethics and genomic editing using the CRISPR-Cas9 technique: Challenges and conflicts. NanoEthics, 11(3), 329-341. https://doi.org/10.1007/s11569-017-0308-1

Liang, P., Xu, Y., Zhang, X., Ding, C., Huang, R., Zhang, Z., ... & Huang, J. (2015). CRISPR-Cas9: A powerful tool for gene editing. BMC Medical Ethics, 16(1), 33-39. https://doi.org/10.1186/s12910-015-0027-7

Shkomova, L. V., Jamal, M., Espinosa, J. E., & Hernández-Hernández, O. (2020). The challenge of CRISPR-Cas toward bioethics. Frontiers in Bioengineering and Biotechnology, 8, 159. https://doi.org/10.3389/fbioe.2020.00159

Regalado, A. (2019). A biographer and a bioethicist take on the CRISPR revolution. Nature. Retrieved from https://www.nature.com/articles/d41586-019-01741-2

6. Quantum AI Safety and Cognitive Capital

Alexeev, Y., et al. (2024). Artificial intelligence for quantum computing. ArXiv Preprint. Retrieved from https://arxiv.org/abs/2411.09131

Lavazza, A. (2018). Cognitive enhancement through genetic editing: A new frontier to explore (and to regulate)? Journal of Bioethical Inquiry, 15(2), 203–217. https://doi.org/10.1007/s41465-018-0104-1

Boretti, A. (2024). Technical, economic, and societal risks in the progress of artificial intelligence driven quantum technologies. Quantum Computing

Perspectives, 3(1), 45–62. https://doi.org/10.1007/s44163-024-00171-y

7. Broadening Perspectives

Rohn, TT et al. "Genetic Modulation of the Htr2a Gene Reduces Anxiety-Related Behavior in Mice", PNAS Nexus, 2023

Kardas P, Lewek P, Matyjaszczyk M. Determinants of patient adherence: A review of systematic reviews. Front Pharmacol. 2013;4:91. doi:10.3389/fphar.2013.00091.

Centers for Disease Control and Prevention. Non-adherence to treatment and its impact on chronic disease outcomes. MMWR Morb Mortal Wkly Rep. 2017;66(45):1238-1242. Available at: https://www.cdc.gov/mmwr/volumes/66/wr/mm6645a2.htm.

Varman DR, Jayanthi LD, Ramamoorthy S. Kappa opioid receptor-mediated differential regulation of serotonin and dopamine transporters in mood and substance use disorder. In: The Kappa Opioid Receptor. Springer; 2021:97-112. doi:10.1007/164_2021_499.

Meier IM, Eikemo M, Leknes S. The role of mu-opioids for reward and threat processing in humans: implications for addiction. Curr Addict Rep. 2021;8(3):365-374. doi:10.1007/s40429-021-00366-8.

Part Two: Individual

8. The Future of RNA Therapies for Enhancing Cognitive Health

Allen, T. M., & Cullis, P. R. (2013). Liposomal drug delivery systems: From concept to clinical applications. *Advanced Drug Delivery Reviews,* 65(1), 36-48. https://doi.org/10.1016/j.addr.2012.09.037

Baden, L. R., El Sahly, H. M., Essink, B., Kotloff, K., Frey, S., Novak, R., ... & Zaks, T. (2021). Efficacy and safety of the mRNA-1273 SARS-CoV-2 vaccine. New England Journal of Medicine, 384(5), 403-416. https://doi.org/10.1056/NEJMoa2035389

Begley, D. J. (2004). Delivery of therapeutic agents to the central nervous system: The problems and the possibilities. *Pharmacology & Therapeutics,* 104(1), 29-45. https://doi.org/10.1016/j.pharmthera.2004.08.001

Jackson, A. L., Burchard, J., Leake, D., Reynolds, A., Schelter, J., Guo, J., ... & Linsley, P. S. (2010). Recognizing and avoiding siRNA off-target effects for target identification and therapeutic application. *Nature Reviews Drug Discovery,* 5(7), 561-565. https://doi.org/10.1038/nrd1878

Sahin, U., Karikó, K., & Türeci, Ö. (2014). mRNA-based therapeutics—developing a new class of drugs. *Nature Reviews Drug Discovery,* 13(10), 759-780. https://doi.org/10.1038/nrd4278

9. The Science of Stress Reduction

Macciardi, F. M., Petronis, A., Van Tol, H. H. M., et al. (1994). Analysis of the D4 dopamine receptor gene variant in an Italian family study. *Journal of Neurogenetics,* 10(2), 101-114.

Nicoraș, V., et al. (2024). Epigenetic and coping mechanisms of stress in affective disorders: A scoping review. *Medicina,* 60(5), 709. https://doi.org/10.3390/medicina60050709

Smith, J., et al. (2024). Stress-induced epigenetic modifications and mental health. *Journal of Neurogenetics*, 29(4), 203-217.

Williams, E., et al. (2024). Genetic and epigenetic contributions to stress responses. *Neurobiology of Stress,* 10(2), 101-114.

10. Memory Decline and Cognitive Impairment

Dinius, C. J., et al.., (2023). Cognitive interventions for memory and psychological well-being in aging and dementias. *Frontiers in Psychology,* 14.

Gould, K. L., Ornish, D., Scherwitz, L., et al.. (2023). Effects of intensive lifestyle changes on the progression of mild cognitive impairment or early dementia due to Alzheimer's disease: a randomized, controlled clinical trial. *Alzheimer's Research & Therapy,* 15.

Kandel, E. R., Schwartz, J. H., Jessell, T. M., Siegelbaum, S. A., & Hudspeth, A. J. (2021). Principles of neural science. McGraw Hill.

McKhann, G. M., Knopman, D. S., Chertkow, H., et al.. (2023). Diagnosis of dementia due to Alzheimer's disease: recommendations from the National Institute on Aging-Alzheimer's Association workgroups. *Alzheimer's Dementia,* 19(7).

Morris, M. C., Evans, D. A., Bienias, J. L., et al.. (2023). Healthy lifestyle and the risk of Alzheimer dementia: Findings from longitudinal studies. *Neurology*, 19(4), 374-383.

Rohn, T., et al., Intranasal Delivery of shRNA to Knockdown the 5-HT2a receptor Enhances Memory and Alleviates Anxiety, (2023)

Tang, H. Y., Vitiello, M. V., Perlis, M., et al.. (2023). Mild Cognitive Impairment in Relation to Alzheimer's disease: An Investigation of Principles, Classifications, Ethics, and Problems. *Neuroethics,* 16(3).

12. Next-Generation Personal Change By J. L. Mee

Brandmeyer, Tracy, et al. "Neurogenetic Regulation of Memory and Anxiety Disorders." *Journal of Neurogenetics,* vol. 32, no. 4, 2023, pp. 415-430.

Rohn, Troy T., et al. "Therapeutic Implications of Targeting 5-HT2a Receptor Hyperactivity in Cognitive Decline." *Journal of Neuroscience Research,* vol. 41, no. 5, 2023, pp. 652-668.

13. Cognitive Resilience Continuum

Cognitive functioning is mapped along a spectrum of durable and measurable states of cognition and the baseline cognitive function for each state.

Part Three: Brain

14. Long-Lasting Genetic Therapeutics for Debilitating Neurological Conditions

By Dean Radin Ph.D.

Anagha K, Shihabudheen P, Uvais NA. Side Effect Profiles of Selective Serotonin Reuptake Inhibitors: A Cross-Sectional Study in a Naturalistic Setting. Prim Care Companion CNS Disord. 2021 Jul 29;23(4):20m02747. doi: 10.4088/PCC.20m02747. PMID: 34324797.

Anderson ND. State of the science on mild cognitive impairment (MCI). CNS Spectr. 2019 Feb;24(1):78-87. doi: 10.1017/S1092852918001347. Epub 2019 Jan 17. PMID: 30651152.

Garcia-Romeu A, Darcy S, Jackson H, White T, Rosenberg P. Psychedelics as Novel Therapeutics in Alzheimer's Disease: Rationale and Potential Mechanisms. Curr Top Behav Neurosci. 2022;56:287-317. doi: 10.1007/7854_2021_267. PMID: 34734390.

Hakimi Y, Petitpain N, Pinzani V, Montastruc JL, Bagheri H. Paradoxical adverse drug reactions: descriptive analysis of French reports. Eur J Clin Pharmacol. 2020 Aug;76(8):1169-1174. doi: 10.1007/s00228-020-02892-2. Epub 2020 May 16. PMID: 32418024.

Lewis V, Bonniwell EM, Lanham JK, Ghaffari A, Sheshbaradaran H, Cao AB, Calkins MM, Bautista-Carro MA, Arsenault E, Telfer A, Taghavi-Abkuh FF, Malcolm NJ, El Sayegh F, Abizaid A, Schmid Y, Morton K, Halberstadt AL, Aguilar-Valles A, McCorvy JD. A non-hallucinogenic LSD analog with therapeutic potential for mood disorders. Cell Rep. 2023 Mar 28;42(3):112203. doi: 10.1016/j.celrep.2023.112203. Epub 2023 Mar 6. PMID: 36884348; PMCID: PMC10112881.

Orgeta V, Leung P, Del-Pino-Casado R, Qazi A, Orrell M, Spector AE, Methley AM. Psychological treatments for depression and anxiety in dementia and mild cognitive impairment. Cochrane Database Syst Rev. 2022 Apr 25;4(4):CD009125. doi: 10.1002/14651858.CD009125.pub3. PMID: 35466396; PMCID: PMC9035877.

Raj P, Rauniyar S, Sapkale B. Psychedelic Drugs or Hallucinogens: Exploring Their Medicinal Potential. Cureus. 2023 Nov 13;15(11):e48719. doi: 10.7759/cureus.48719. PMID: 38094517; PMCID: PMC10716812.

Rohn TT, Radin D, Brandmeyer T, Linder BJ, Andriambeloson E, Wagner S, Kehler J, Vasileva A, Wang H, Mee JL, Fallon JH. Genetic modulation of the HTR2A gene reduces anxiety-related behavior in mice. PNAS Nexus. 2023 Jun 20;2(6):pgad170. doi: 10.1093/pnasnexus/pgad170. PMID: 37346271; PMCID: PMC10281383.

van den Berg M, Magaraggia I, Schreiber R, Hillhouse TM, Porter JH. How to account for hallucinations in the interpretation of the antidepressant effects of psychedelics: a translational framework. Psychopharmacology (Berl). 2022 Jun;239(6):1853-1879. doi: 10.1007/s00213-022-06106-8. Epub 2022 Mar 29. PMID: 35348806; PMCID: PMC9166823.15. Hippocampal Neuron Hyperactivity and Longevity

Rohn, T. T., Rissman, R. A., Davis, M. C., Kim, Y. E., & Cotman, C. W. (2002). Caspase-9 activation and caspase cleavage of tau in the Alzheimer's disease brain. *Neurobiology of Disease,* 11(2), 341-354. https://doi.org/10.1016/S0969-9961(02)00005-4

Rohn, T. T., Radin, D., & Brandmeyer, T. (2024). Long-lasting genetic therapeutics for debilitating neurological conditions. *Nature Biopharmdeal,* March 2024.

Rohn, T. T., Head, E., & Cotman, C. W. (2004). Caspase activation, independent of cell death, is required for proper cell dispersal and correct morphology in PC12 cells. *Journal of Neuroscience Research,* 75(3), 353-362. https://doi.org/10.1002/jnr.10845

16. Artificial Intelligence and Genetic Medicine

Ahmed, Zeeshan, et al.. "Artificial Intelligence for Personalized and Predictive Genomics Data Analysis." Frontiers in Genetics, vol. 14, 2023, doi:10.3389/fgene.2023.1162869.

Ellinor, Patrick T., et al.. "Transfer Learning Enables Predictions in Network Biology." Nature, 2023, doi:10.1038/s41586-023-06139-9.

Liu, P., et al.. "Speeding the Diagnosis of Rare Genetic Disorders with the Help of Artificial Intelligence." NIH Director's Blog, 2024.

Vadapalli, Ramesh, et al.. "Artificial Intelligence System Predicts Consequences of Gene Modifications." Medical Xpress, 2023.

17. An Inquiry Into Brain Coherence

Joseph Mee, et al.., *The Role of 5-HT2a Receptors in Brain Coherence and Mental Health Disorders,* (2023)

Fabio Macciardi, et al.., *Therapeutic Innovations Targeting Brain Coherence: The Role of RNA-Based Therapies,* (2023)

Hal Blumenfeld, *The Neurobiology of Consciousness: Cognitive Neuroscience and the Inner Life of the Mind*

18. Innovative Longevity Biomarkers for CNS Disorders

Allen, T. M., & Cullis, P. R. (2013). Liposomal drug delivery systems: From concept to clinical applications. *Advanced Drug Delivery Reviews*, 65(1), 36–48. https://doi.org/10.1016/j.addr.2012.09.037

Collins, F. S., & Varmus, H. (2015). A new initiative on precision medicine. *The New England Journal of Medicine*, 372(9), 793–795. https://doi.org/10.1056/NEJMp1500523

McKee, A. C., Stern, R. A., Nowinski, C. J., Stein, T. D., Alvarez, V. E., Daneshvar, D. H., ... & Cantu, R. C. (2013). The neuropathology of chronic traumatic encephalopathy. *Brain Pathology,* 23(3), 350–364. https://doi.org/10.1111/bpa.12028

Rohn, T. T. (2018). The role of genetic therapies in Alzheimer's disease. *Journal of Alzheimer's disease & Parkinsonism,* 8(3), 439. https://doi.org/10.4172/2161-0460.1000439

Sahin, U., Karikó, K., & Türeci, Ö. (2014). mRNA-based therapeutics—developing a new class of drugs. *Nature Reviews Drug Discovery,* 13(10), 759–780. https://doi.org/10.1038/nrd4278

19. Overcoming the Blood-Brain Barrier

Larsen, J. M., et al.,(2024). Recent advances in delivery through the blood-brain barrier. *Current Topics in Medicinal Chemistry,* 24(9), 1148-1160. https://doi.org/10.2174/1568026614666140329230311

Patel, M. M., & Patel, B. M. (2023). Strategies to improve drug delivery across the blood-brain barrier: A review. *Neuro-Oncology,* 25(1), 45-60. https://doi.org/10.1093/neuonc/noae087

Sweeney, M. D., Zhao, Z., & Montagne, A. (2023). Advances in blood-brain barrier transport: Focus on brain delivery systems. *Journal of Neurochemistry,* 166(1), 70-87. https://doi.org/10.1111/jnc.15573

Part Four: Neuron

21. Precise Genetic Targeting in Neuropsychiatric Disorders

Allen, B. (2024). The promise of explainable AI in digital health for precision medicine: A systematic review. *Journal of Personalized Medicine,* 14(3), 277. https://doi.org/10.3390/jpm14030277

Carlos, A. J., Lee, M., & Siman, R. (2010). Overexpression of Bcl-2 in APP transgenic mice reduces amyloid pathology. *Neurobiology of Aging,* 31(6), 891-897. https://doi.org/10.1016/j.neurobiolaging.2008.07.001

Rohn, T. T., et al.., (2001). 15-Deoxy-Δ12,14-prostaglandin J2, a specific ligand for peroxisome proliferator-activated receptor-γ, induces neuronal apoptosis. *Journal of Neurochemistry,* 78(3), 503-514. https://doi.org/10.1046/j.1471-4159.2001.00445.x

22. How RNA Therapies Target Neurons for Modulation

Allen, T. M., & Cullis, P. R. (2013). Liposomal drug delivery systems: From concept to clinical applications. *Advanced Drug Delivery Reviews,* 65(1), 36–48. https://doi.org/10.1016/j.addr.2012.09.037

Collins, F. S., & Varmus, H. (2015). A new initiative on precision medicine. *The New England Journal of Medicine,* 372(9), 793-795. https://doi.org/10.1056/NEJMp1500523

McKee, A. C., et al. (2013). The neuropathology of chronic traumatic encephalopathy. *Brain Pathology,* 23(3), 350–364. https://doi.org/10.1111/bpa.12028

Sahin, U., Karikó, K., & Türeci, Ö. (2014). mRNA-based therapeutics—developing a new class of drugs. *Nature Reviews Drug Discovery,* 13(10), 759-780. https://doi.org/10.1038/nrd4278

23. Harnessing RNA Therapies to Address Neuroinflammation

Kumar, A., et al.. "Advances in Gene Therapy Approaches Targeting Neuroinflammation in Neurodegenerative Diseases." Gene Therapy, vol. 28, no. 5, 2021, pp. 321-337. *Springer Nature.* doi:10.1038/s41434-021-00256-7.

Morrison, B., et al.. "RNA Interference (RNAi): Harnessing Molecular Mechanisms for Therapeutic Applications in Neuroinflammation." Genetics and Molecular Research, vol. 20, no. 1, 2021, pp. 1-15. *Fundacão de Pesquisas Científicas de Ribeirão Preto.* doi:10.4238/gmr18919.

Sarkar, A., et al.. "A New Perspective on Depression and Neuroinflammation: Non-Coding RNA." Neurobiology of Disease, vol. 148, 2021, article 105179. *Elsevier.* doi:10.1016/j.nbd.2020.105179.

Yang, Z., et al.. "Emerging Role of Non-Coding RNAs in Neuroinflammation Mediated by Microglia and Astrocytes." *Journal of Neuroinflammation,* vol. 18, 2021, pp. 1-20. BioMed Central. doi:10.1186/s12974-021-02139-4

24. Synthetic Biology: A New Era in Biotechnology
By J. L. Mee

Amy Webb, et al.., *The Genesis Machine: Our Quest to Rewrite Life in the Age of Synthetic Biology,* (2022)

Yuval Harari, *Homo Deus: A Brief History of Tomorrow,* (2015)

Ugur Sahin, et al.., *mRNA-based therapeutics—developing a new class of drugs,* (2014)

Part Five: Pillars of Innovation

26. Scientific Review

27. Paradigm Shift

Article Title: Intranasal Delivery of shRNA to Knockdown the 5-HT2aReceptor Enhances Memory and Alleviates Anxiety.

28. Blueprint for Tomorrow's Medicine

Article Title: Treatment with shRNA to knock down the 5-HT2a receptor improves memory in vivo and decreases excitability in primary cortical neurons.

29. The Dawn Of Precision Medicine For Mental Health

Article Title: Genetic Modulation of the HTR2A Gene Reduces Anxiety-Related Behavior in Mice.

iii. Brain Networks

Overview of Brain Networks and RNA Neuromodulation

The human brain contains approximately 86 billion neurons forming intricate networks that collaborate to regulate motor control, sensory processing, attention, memory, emotional regulation, decision-making, and reward processing. These networks operate not as isolated systems but as a dynamic, interconnected orchestra whose coordinated activity gives rise to the full spectrum of human cognition and behavior.

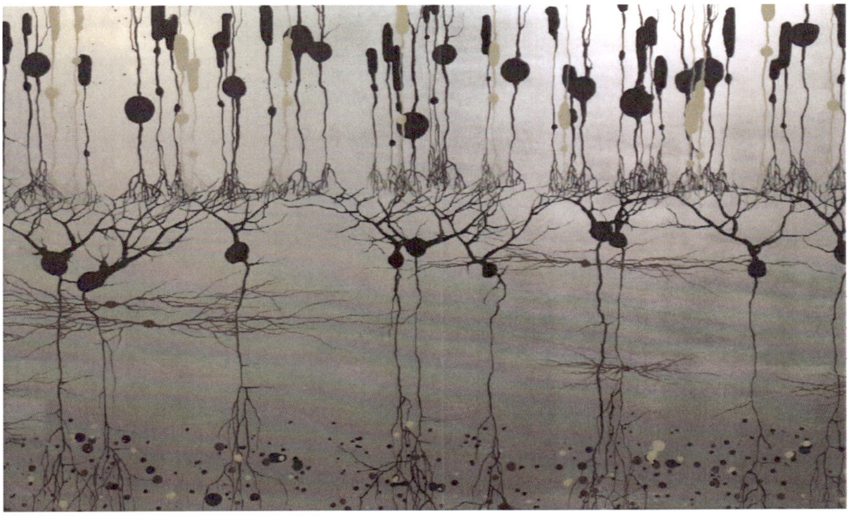

Retina, Greg Dunn

As mammals evolved, networks supporting spatial navigation and memory (hippocampal network) and emotional processing (limbic system) expanded, facilitating more sophisticated environmental interactions and social behaviors. The extensive development of prefrontal networks represents one of the most recent evolutionary innovations, particularly prominent in primates and reaching its apex in humans. This evolutionary layering explains why primitive emotional responses sometimes override rational decision-making—our newer networks must regulate systems that evolved millions of years earlier.

The default mode network (DMN) appears to be particularly expanded in humans compared to other primates, suggesting its importance in distinctly human capabilities such as self-reflection, mental time travel, and abstract thinking. This evolutionary perspective helps explain why certain networks are more vulnerable to dysfunction than others, informing how RNA neuromodulators might be targeted to address specific disorders while preserving more evolutionarily conserved functions.

Network Development Across the Lifespan

Brain networks follow distinct developmental trajectories from infancy through adulthood. During early development, sensory and motor networks mature first, establishing basic processing capabilities. The visual network, for instance, undergoes rapid development in the first year of life as infants learn to process increasingly complex visual information.

Networks supporting attention and cognitive control, such as the frontoparietal and dorsal attention networks, undergo significant maturation during childhood and adolescence, paralleling improvements in self-regulation and sustained attention. The prefrontal cortex network continues developing into early adulthood, explaining why executive functions like impulse control and long-term planning improve into the mid-20s.

As we age, some networks maintain their integrity while others show decline. The default mode network often shows reduced connectivity in older adults, potentially contributing to changes in memory and self-referential processing. Understanding these developmental trajectories provides crucial context for timing interventions with RNA neuromodulators, as targeting specific networks may be more effective at certain developmental stages.

Studying Brain Networks: Imaging and Analysis Techniques

Our understanding of brain networks has been revolutionized by advances in neuroimaging and analysis techniques. Functional magnetic resonance imaging (fMRI) measures blood oxygen level-dependent (BOLD) signals that correlate with neural activity, allowing researchers to identify regions that activate together during rest or specific tasks. Diffusion tensor imaging (DTI) reveals the structural connections—white matter tracts—that enable communication between distant brain regions.

Electroencephalography (EEG) and magnetoencephalography (MEG) capture the electrical and magnetic signals generated by neural activity with millisecond precision, providing insights into the temporal dynamics of network interaction. Graph theory and network neuroscience approaches allow researchers to quantify properties like network efficiency, modularity, and hub centrality, revealing how information flows through the brain's complex architecture.

These methods have transformed our conceptualization of brain function from a focus on isolated regions to an appreciation of distributed, interacting networks. This network perspective provides a more sophisticated framework for understanding both normal cognition and the disruptions that occur in neuropsychiatric disorders.

Network Oscillations: The Brain's Communication Frequencies

Brain networks coordinate their activity through oscillations—rhythmic patterns of neural activity occurring at different frequencies. These oscillations serve as the brain's communication channels, enabling synchronization both within and between networks.

Theta oscillations (4-8 Hz), generated in the hippocampus, play crucial roles in spatial navigation and memory formation. Alpha oscillations (8-12 Hz) in visual and sensorimotor regions help suppress irrelevant information. Beta oscillations (13-30 Hz) coordinate motor planning and execution. Gamma oscillations (30-100 Hz) facilitate local processing and feature binding across the cortex.

Different networks favor different oscillatory patterns. The default mode network, for instance, shows prominent activity in slow frequencies, while attentional networks engage faster beta and gamma oscillations during focused concentration. Disruptions in these rhythmic patterns are associated with various disorders—excessive beta activity in the corticofugal network characterizes Parkinson's disease, while abnormal gamma synchronization in frontoparietal regions appears in schizophrenia.

RNA neuromodulators might be specifically designed to normalize these oscillatory patterns, targeting the specific receptor systems and neural populations that generate and regulate brain rhythms. This approach represents a more sophisticated intervention than traditional pharmacology, which often broadly affects neurotransmitter systems throughout the brain.

Corticofugal Network

The corticofugal network consists of neural pathways descending from the cerebral cortex to subcortical structures such as the thalamus, brainstem, and spinal cord. This network is responsible for motor control and sensory processing, coordinating the brain's top-down influence on bodily movement and sensory perception.

Evolutionary Perspective: The corticofugal network has ancient evolutionary origins, with basic descending motor pathways present even in primitive vertebrates. However, the expansion of direct connections between motor cortex and spinal motor neurons (the corticospinal tract) represents a more recent evolutionary development, particularly pronounced in primates and humans. This evolutionary innovation enables the fine motor control necessary for tool use and complex manipulation.

Network Interactions: The corticofugal network communicates extensively with the motor cortex, thalamic networks, and spinal cord circuits. It synchronizes sensory information with motor output and regulates movement and reflexes. This network enhances the brain's capacity to govern muscular movements by transferring signals from the cortex to subcortical areas. It also modulates sensory information as it travels to the brain, influencing the body's reaction to external stimuli.

Clinical Relevance: Disruptions in the corticofugal network are implicated in movement disorders such as Parkinson's disease, where abnormal activity in basal ganglia circuits affects descending motor control. Stroke affecting corticofugal pathways can result in paralysis or motor weakness, depending on specific tracts damaged.

RNA Neuromodulator Applications: RNA neuromodulators targeting the corticofugal network might focus on normalizing abnormal activity patterns in conditions like Parkinson's disease. By regulating the expression of specific receptors or ion channels in corticofugal neurons, these interventions could potentially restore more normal patterns of motor control without the side effects associated with traditional medications like levodopa.

Default Mode Network

The Default Mode Network (DMN) includes the medial prefrontal cortex, posterior cingulate cortex, precuneus, and angular gyrus. It shows highest activity during rest and internal reflection, making it unique among brain networks that typically activate in response to external demands.

Evolutionary Development: While simple forms of the DMN exist in other mammals, its elaborate development appears to be a particularly human trait. Comparative studies show that the human DMN has expanded connectivity and more distributed hubs compared to non-human primates, potentially supporting uniquely human capabilities like autobiographical memory, theory of mind, and future planning.

Developmental Trajectory: The DMN undergoes significant maturation from childhood through adolescence. In infants and young children, the network exists but shows less coordinated activity. As development progresses, DMN connectivity strengthens and becomes more integrated, paralleling improvements in self-awareness and abstract thinking.

Network Oscillations: The DMN typically operates at lower frequencies (1-8 Hz), particularly in the theta range, allowing for integration of information across distant brain regions. These slower oscillations may facilitate the network's role in memory consolidation and reflective thought.

The DMN interacts dynamically with the Frontoparietal Network and the Salience Network, maintaining equilibrium between internal cognitive processes and exterior goal-directed tasks. This facilitates transitions between introspective and outwardly focused mental states. The network exhibits its highest level of activity when the brain is not engaged with external stimuli, such as during daydreaming or introspection. It is essential for preserving a sense of identity and incorporating previous experiences into memory.

Clinical Significance: Disruptions in DMN activity and connectivity are associated with psychiatric and neurological conditions. In depression, the DMN often shows hyperactivity, potentially contributing to rumination and self-focused negative thinking. Alzheimer's disease features prominent DMN dysfunction, with reduced connectivity appearing early in the disease process. Schizophrenia is characterized by altered DMN connectivity patterns that may contribute to disturbances in self-perception and reality testing.

RNA Neuromodulator Potential: RNA neuromodulators targeting the DMN might focus on normalizing network hyperactivity in depression or strengthening declining connectivity in early Alzheimer's disease. By targeting specific receptor systems like 5-HT2a receptors in key DMN hubs, these interventions could potentially rebalance network activity precisely.

Dorsal Attention Network

The Dorsal Attention Network includes the intraparietal sulcus and frontal eye fields. It activates when attention is voluntarily directed to specific stimuli or tasks, playing a crucial role in top-down attentional control.

Evolutionary Context: Attentional networks emerged as creatures needed to prioritize certain environmental stimuli over others for survival. While basic forms of the dorsal attention system exist in many vertebrates, the extensive frontal-parietal connections in primates and humans allow for more sophisticated and sustained attentional control.

Network Plasticity: The Dorsal Attention Network shows remarkable plasticity, strengthening with training and practice. Studies of experienced meditators reveal enhanced connectivity and activity in this network, corresponding to improved attentional control. This plasticity suggests the potential for targeted interventions to enhance attention in conditions characterized by attentional deficits.

The Dorsal Attention Network is responsible for regulating goal-directed attention and visuospatial processes, facilitating the ability to concentrate on pertinent inputs. The network interacts with the Ventral Attention Network and the Frontoparietal Network, collaborating to sustain attention, adjust focus as necessary, and merge cognitive activities with sensory information.

This network assists the brain in concentrating on tasks that demand prolonged attention, such as reading or navigating in physical surroundings. It effectively eliminates distractions and enables focused attention on crucial activities.

Clinical Relevance: Malfunctioning in this network can result in impairments in attention and spatial cognition, frequently observed in disorders such as ADHD and dementia. In ADHD, the Dorsal Attention Network often shows reduced activation during sustained attention tasks and altered connectivity with other networks involved in cognitive control.

RNA Neuromodulator Applications: RNA neuromodulators targeting the Dorsal Attention Network might focus on enhancing network efficiency in conditions like ADHD. By modulating the expression of dopamine or norepinephrine receptors in key network nodes, these interventions could potentially improve attentional capabilities without the side effects associated with traditional stimulant medications.

Frontoparietal Network

The Frontoparietal Network encompasses the dorsolateral prefrontal cortex, inferior parietal lobule, and parts of the cingulate cortex. It plays a central role in cognitive control and working memory—the ability to hold and manipulate information in mind.

Developmental Significance: The Frontoparietal Network undergoes protracted development throughout childhood and adolescence, with significant maturation occurring in the teenage years and early adulthood. This extended developmental trajectory parallels improvements in executive functions like cognitive flexibility, planning, and inhibitory control.

Network Oscillations: The Frontoparietal Network typically operates in the beta (13-30 Hz) and gamma (30-100 Hz) frequency ranges during active cognitive processing. These faster oscillations facilitate rapid updating of information in working memory and flexible shifts between different task demands.

The Frontoparietal Network plays a major role in high-level cognitive tasks, including working memory, decision-making, and problem-solving. The network interacts with the Default Mode Network, the Dorsal Attention Network, and the Salience Network. This interaction allows for the integration of information from different networks, enabling the individual to direct their behavior and adapt to cognitive demands that may change over time.

This network plays a crucial role in performing intricate cognitive activities, such as strategizing, logical thinking, and adjusting to unfamiliar circumstances. It combines sensory input with past experiences to produce well-informed choices.

Cultural Variations: Interestingly, while the basic architecture of the Frontoparietal Network appears consistent across cultures, its engagement during specific cognitive tasks shows some cultural variation. Studies comparing East Asian and Western participants have found differences in how this network engages during certain reasoning tasks, potentially reflecting cultural influences on cognitive styles.

Clinical Significance: Disruption in this network is linked to cognitive impairments in illnesses such as schizophrenia, Alzheimer's disease, and other neuropsychiatric disorders. In schizophrenia, reduced Frontoparietal Network connectivity correlates with deficits in working memory and cognitive control.

RNA Neuromodulator Potential: RNA neuromodulators targeting the Frontoparietal Network might focus on restoring normal network function in conditions like schizophrenia. By targeting glutamate receptor expression in prefrontal regions, these interventions could potentially improve cognitive control without the broad effects of current antipsychotic medications.

Hippocampal Network

The Hippocampal Network centers around the hippocampus and includes connections with the entorhinal cortex, parahippocampal gyrus, and retrosplenial cortex. This network is fundamental for episodic memory formation and spatial navigation.

Evolutionary Perspective: The hippocampus is an ancient structure, present in all mammals and with homologous regions in other vertebrates. Its role in spatial navigation appears to be evolutionarily conserved, while its involvement in episodic memory represents an elaboration particularly developed in humans. This evolutionary history explains why basic spatial learning can remain intact even when episodic memory is impaired.

Network Plasticity: The Hippocampal Network exhibits remarkable plasticity throughout life. The hippocampus is one of the few brain regions capable of neurogenesis in adulthood, though at a much lower rate than during development. This capacity for generating new neurons may contribute to the network's role in forming new memories and adapting to changing environments.

The Hippocampal Network plays a crucial role in memory consolidation, spatial orientation, and emotional modulation. The network interacts with the Limbic System, the Default Mode Network, and the Prefrontal Cortex Network. It organizes the processing of memory and emotional reactions, integrating spatial and emotional contexts into the preservation of memories.

This network is responsible for the processing and consolidation of information from short-term memory into long-term storage. It also aids in orienting oneself within surroundings by constructing cognitive representations of the environment. The hippocampus plays a crucial role in emotional regulation by closely collaborating with the amygdala to analyze and react to emotional events.

Clinical Relevance: Excessive activity in this network is frequently observed in neurodegenerative disorders such as Alzheimer's disease, as well as in epilepsy and schizophrenia, where there is impairment in memory and emotional control. Alzheimer's disease typically begins with hippocampal dysfunction, explaining why memory loss is often the first symptom. Temporal lobe epilepsy frequently involves abnormal hippocampal activity, sometimes resulting in memory disturbances between seizures.

RNA Neuromodulator Applications: RNA neuromodulators targeting the Hippocampal Network might focus on reducing hyperactivity in conditions like epilepsy or strengthening connectivity in early

Alzheimer's disease. By modulating the expression of specific glutamate receptor subunits or potassium channels in hippocampal neurons, these interventions could potentially normalize network function with fewer side effects than current medications.

Limbic System

The Limbic System includes the amygdala, hippocampus, hypothalamus, anterior cingulate cortex, and parts of the orbitofrontal cortex. Often called the "emotional brain," this interconnected network processes emotions, drives, and associated memories.

Evolutionary Significance: The limbic system represents one of the more evolutionarily ancient parts of the mammalian brain, developing to process threat, reward, and social information critical for survival. The basic architecture of structures like the amygdala appears remarkably conserved across mammals, highlighting their fundamental importance in emotion processing.

Developmental Trajectory: Different components of the limbic system mature at varying rates. The amygdala develops relatively early, explaining why basic emotional responses are present from infancy. In contrast, connections between the limbic system and prefrontal regions continue to develop into adolescence and early adulthood, paralleling improvements in emotional regulation.

The Limbic System is responsible for regulating emotions, memory, and motivation through the involvement of components such as the hippocampus and amygdala. It interacts with the Hippocampal Network, the Salience Network, and the Prefrontal Cortex Network, processing emotional cues, guiding reactions based on prior experiences, and connecting emotional and cognitive processing.

This intricate network is responsible for regulating emotional reactions, memory, and motivational states. The amygdala has a pivotal function in the processing of emotions and the formation of memories associated with those feelings. This system is essential for survival, since it directs actions such as fear responses and the pursuit of pleasure.

Cultural Variations: While basic emotional processing appears universal, cultural factors can influence limbic system function. Cross-cultural neuroimaging studies have found differences in amygdala responsivity to various emotional stimuli that correlate with cultural values regarding emotional expression and regulation.

Clinical Significance: The limbic system is a crucial target for interventions addressing emotional and cognitive disorders. Anxiety disorders often involve amygdala hyperactivity, while depression may feature disrupted connectivity between limbic structures and regulatory prefrontal regions. Post-traumatic stress disorder (PTSD) is characterized by limbic dysregulation, particularly in threat processing circuits involving the amygdala.

RNA Neuromodulator Potential: RNA neuromodulators targeting the limbic system could offer precise interventions for conditions like anxiety and PTSD. By targeting specific receptor systems in the amygdala, these therapies might normalize emotional processing without affecting cognition or motivation. This precision represents a significant advance over current anxiolytics, which often cause sedation or cognitive blunting.

Prefrontal Cortex Network

The Prefrontal Cortex Network includes the dorsolateral, ventromedial, and orbitofrontal regions of the prefrontal cortex, along with their connections to other cortical and subcortical structures. This network sits at the apex of the cognitive hierarchy, coordinating and regulating other networks.

Evolutionary Context: The prefrontal cortex is particularly expanded in humans compared to other primates, especially its dorsolateral portions. This evolutionary expansion parallels the development of uniquely human capabilities like complex planning, abstract reasoning, and moral judgment. The prefrontal cortex's late evolution makes it both our most sophisticated neural asset and one of the most vulnerable to dysfunction.

Developmental Timeline: The prefrontal cortex has the most protracted developmental course of any brain region, continuing to mature into the mid-twenties. Myelination of prefrontal circuits and pruning of excess synapses proceeds throughout adolescence, explaining improvements in impulse control and decision-making during this period. This extended development creates both vulnerability and opportunity—while the developing prefrontal cortex may be susceptible to stressors, it also provides a window for intervention.

The primary role of the Prefrontal Cortex Network is to regulate executive processes, such as decision-making, impulse control, and working memory. It interacts with the Frontoparietal Network, the Limbic System, and the Hippocampal Network, overseeing complex cognitive processes and integrating emotional and memory-related information to facilitate decision-making.

This network is responsible for regulating intricate activities, such as strategic planning, social interactions, and self-regulation. It aids in evaluating the repercussions of actions and making decisions based on long-term objectives rather than impulsive urges.

Clinical Relevance: Malfunctioning within this network can result in difficulties with focus, self-restraint, and emotional regulation, hence leading to mental health disorders such as anxiety, depression, and ADHD. Prefrontal dysfunction is evident in numerous psychiatric conditions—reduced prefrontal activation during cognitive tasks characterizes ADHD, while depression often features altered activity in ventromedial prefrontal regions involved in emotion regulation.

RNA Neuromodulator Applications: RNA neuromodulators targeting the Prefrontal Cortex Network might focus on enhancing network function in conditions like ADHD or depression. By modulating the expression of dopamine receptors in dorsolateral regions or serotonin receptors in ventromedial areas, these interventions could potentially improve specific aspects of prefrontal function without broadly affecting neurotransmitter systems throughout the brain.

Salience Network

The Salience Network includes the anterior insula and dorsal anterior cingulate cortex. This network acts as a dynamic switch, identifying significant stimuli from the constant stream of internal and external information and directing other networks to respond.

Network Oscillations: The Salience Network typically operates across multiple frequency bands, with prominent activity in the theta (4-8 Hz) range during emotional processing and in faster gamma (30-100 Hz) frequencies when detecting salient external stimuli. This frequency flexibility may support the network's role in coordinating responses across different timescales.

Network Plasticity: The Salience Network shows considerable plasticity in response to experience. In individuals with chronic pain, this network often becomes hypersensitive to pain signals, potentially contributing to pain chronification. Conversely, mindfulness meditation training can normalize Salience Network responses to emotional stimuli, suggesting potential for targeted interventions to modulate its function.

The Salience Network is responsible for detecting and filtering significant inputs, and coordinating appropriate reactions to critical changes in the environment. It interacts with the Default Mode Network, the Dorsal Attention Network, and the Limbic System, facilitating the shifting of focus between internal and external stimuli and prioritizing crucial information for prompt reaction.

This network plays a role in detecting and reacting to the most significant stimuli in the surroundings, whether external (such as a loud noise) or internal (such as a powerful emotion). The brain's ability to shift attention between tasks and process emotions is heavily reliant on this function.

Clinical Significance: Disruption in this network is associated with conditions such as depression, anxiety, and schizophrenia, where the capacity to prioritize and react effectively to inputs is compromised. In anxiety disorders, the Salience Network often shows hyperresponsivity to potential threat cues, contributing to heightened vigilance and worry. In addiction, this network becomes sensitized to drug-related stimuli, triggering craving and relapse.

RNA Neuromodulator Potential: RNA neuromodulators targeting the Salience Network might focus on normalizing its response thresholds in conditions like anxiety or addiction. By modulating the expression of specific receptor systems in the anterior insula or anterior cingulate, these interventions could potentially recalibrate the network's sensitivity without broadly affecting arousal or attention.

Somatomotor Network

The Somatomotor Network encompasses the primary motor cortex, primary somatosensory cortex, and associated subcortical structures like the cerebellum and basal ganglia. This network processes bodily sensations and controls voluntary movement.

Evolutionary Perspective: The somatomotor system has ancient evolutionary origins, with somatotopic organization (the mapping of body parts to specific brain areas) conserved across vertebrates. However, the direct corticospinal connections enabling fine motor control have expanded significantly in primates and humans, allowing for more precise movements of the hands and fingers.

Network Oscillations: The Somatomotor Network exhibits prominent oscillations in the beta frequency range (13-30 Hz) during motor planning and execution. When movement begins, beta power typically decreases in a phenomenon known as "beta desynchronization," followed by a "beta rebound" after movement completion. Disruptions to these rhythmic patterns characterize movement disorders like Parkinson's disease.

The Somatomotor Network is responsible for processing sensory information and regulating voluntary motor functions, combining sensory inputs with motor instructions. It interacts with the Corticofugal Network, the Motor Cortex, and the Thalamic Networks, facilitating seamless and synchronized motions by connecting sensory input with motor output.

This network combines sensory information from the body with motor orders from the brain in order to generate synchronized and intentional motions. The central nervous system is accountable for a wide range of functions, ranging from basic reflexes to intricate activities like walking or playing a musical instrument.*Clinical Relevance:* Disruption in this network can lead to motor impairments, such as tremors or reduced coordination, and changes in sensory processing, often observed in disorders such as Parkinson's disease and stroke. After stroke affecting motor regions, the somatomotor network typically undergoes reorganization, with adjacent brain areas often assuming functions of the damaged tissue.

RNA Neuromodulator Applications: RNA neuromodulators targeting the Somatomotor Network might focus on facilitating network reorganization after stroke or normalizing abnormal activity patterns in movement disorders. By modulating the expression of specific receptors or ion channels in motor neurons, these interventions could potentially enhance recovery or restore more normal movement patterns.

Ventral Attention Network

The Ventral Attention Network includes the temporoparietal junction and ventral frontal cortex, particularly in the right hemisphere. While the Dorsal Attention Network supports voluntary, goal-directed attention, the Ventral Attention Network mediates stimulus-driven attention—our automatic orientation to unexpected or salient stimuli.

Developmental Trajectory: The Ventral Attention Network shows significant maturation during childhood, with increasing specialization for detecting behaviorally relevant unexpected stimuli. This development parallels improvements in children's ability to reorient attention when necessary, a crucial skill for adaptable behavior in changing environments.

Network Asymmetry: The Ventral Attention Network shows pronounced right hemisphere lateralization, particularly for detecting novel or unexpected stimuli. This asymmetry appears to be a conserved feature across individuals and may represent an evolutionary adaptation for rapid detection of potential threats.

The Ventral Attention Network is primarily responsible for recognizing and promptly responding to unforeseen stimuli, as well as sustaining attention on crucial activities. It interacts with the Dorsal Attention Network, the Salience Network, and the Frontoparietal Network to regulate attentional changes and sustain task-oriented concentration, especially in dynamic settings.

This network plays a crucial role in responding to abrupt alterations in the surroundings, such as an unforeseen sound or motion. It facilitates rapid redirection of attention by the brain to effectively respond to novel or significant stimuli, enabling us to adjust to alterations in our environment.

Clinical Significance: Disruption in this network might result in challenges in sustaining focus and reacting suitably to alterations, frequently observed in conditions such as ADHD and autism spectrum disorder. In ADHD, the Ventral Attention Network often shows reduced activity during tasks requiring reorientation to relevant stimuli. In spatial neglect following right hemisphere stroke, damage to this network can result in failure to attend to stimuli in the left visual field.

RNA Neuromodulator Potential: RNA neuromodulators targeting the Ventral Attention Network might focus on enhancing network function in conditions like ADHD or neglect. By modulating the expression of specific receptors in temporoparietal regions, these interventions could potentially improve attentional reorienting without broadly affecting arousal or vigilance.

Visual Network

The Visual Network encompasses the primary and association visual cortices in the occipital lobe, along with visual processing regions extending into temporal and parietal lobes. This network processes visual information through hierarchical pathways that extract increasingly complex features from the visual scene.

Evolutionary Context: Vision is the dominant sense in primates, and the visual system shows remarkable evolutionary conservation across species. However, primate visual cortex has expanded significantly, particularly in areas processing color, form, and motion—adaptations that supported arboreal lifestyles. In humans, regions supporting object recognition and reading represent further specializations.

Developmental Plasticity: The Visual Network exhibits pronounced plasticity during early development, with critical periods for the establishment of basic visual functions like depth perception and binocular vision. However, recent research has revealed surprising adaptability even in adulthood, with perceptual learning able to enhance visual processing throughout life.

The Visual Network is responsible for processing visual information and combining it with other sensory inputs to build a cohesive image of the surrounding world. It interacts with the Dorsal Attention Network, the Somatomotor Network, and the Frontoparietal Network, combining visual perception with attention, movement, and cognitive processing to navigate the surroundings.

This network handles the processing of the immense quantity of visual data that humans constantly receive, including forms, colors, and motion. It combines this information with other sensory inputs to facilitate our comprehension and engagement with our environment.

Clinical Significance: Malfunctioning within this network can result in visual perception impairments, such as challenges in identifying objects or difficulty with spatial orientation, which are frequently found in disorders like visual agnosia and specific forms of dementia. Visual network dysfunction can also contribute to conditions like dyslexia, where abnormal processing in the visual word form area may impair reading.

RNA Neuromodulator Applications: RNA neuromodulators targeting the Visual Network might focus on enhancing network function in conditions with visual processing deficits. By modulating the expression of specific receptors in visual cortex, these interventions could potentially improve visual processing without affecting other sensory or cognitive functions.

Ventral Striatum Network

The Ventral Striatum Network centers around the nucleus accumbens and includes connections with the ventral tegmental area, orbitofrontal cortex, amygdala, and hippocampus. This network forms the core of the brain's reward system, processing incentive value and motivating approach behaviors.

Evolutionary Significance: The reward system has ancient evolutionary origins, with homologous circuits present even in primitive vertebrates. This conservation highlights its fundamental importance in guiding behavior toward beneficial resources and opportunities. While the basic architecture is conserved, human reward circuits show expanded connectivity with prefrontal regions, allowing for more complex reward processing and self-regulation.

Developmental Changes: The Ventral Striatum Network undergoes significant changes during adolescence, with heightened responsivity to rewards compared to childhood or adulthood. This developmental pattern may encourage exploration and learning but can also contribute to risk-taking behavior during this period.

The Ventral Striatum Network is a component of the brain's reward system and is responsible for functions such as motivation, reinforcement learning, and the processing of rewards and pleasure. This network interacts with the Limbic System, the Prefrontal Cortex Network, and the Dopaminergic Pathways, regulating reward-based behavior and decision-making by linking emotional reactions with actions aimed at achieving certain goals.

This network plays a vital role in driving desire and the pursuit of rewards. The brain processes pleasurable experiences and strengthens actions that result in favorable results, such as eating or socializing. The network is intricately linked to the brain's dopamine system, which governs sensations of reward and pleasure.

Cultural Variations: While the basic function of the reward system appears universal, cultural factors can influence what stimuli activate this network. Neuroimaging studies have found cultural differences in ventral striatum responses to various rewards, reflecting cultural values and learning histories.

Clinical Relevance: The disruption of this network is associated with addiction, depression, and other mood disorders, when the equilibrium of reward processing is compromised. In addiction, drugs of abuse hijack this network by directly increasing dopamine release, creating abnormally strong reinforcement that drives compulsive drug-seeking. In depression, reduced ventral striatum activity may contribute to anhedonia—the inability to experience pleasure from previously enjoyable activities.

RNA Neuromodulator Potential: RNA neuromodulators targeting the Ventral Striatum Network might focus on restoring normal reward responsivity in conditions like depression or recalibrating abnormal responses in addiction. By modulating the expression of specific dopamine receptor subtypes in the nucleus accumbens, these interventions could potentially normalize reward processing more precisely than current medications that broadly affect dopaminergic transmission.

Network Integration and Cross-Talk

While we've examined individual networks, the brain's remarkable capabilities emerge from their coordinated interaction. Networks don't function in isolation but engage in constant cross-talk, with information flowing between systems through both direct structural connections and synchronization of oscillatory activity.

The Salience Network often serves as a dynamic switch between the Default Mode Network and Frontoparietal Network, determining whether attention should be directed internally or externally. The Hippocampal Network coordinates with the Ventral Striatum

Network during reward-based learning, linking contextual information with reinforcement signals. The Prefrontal Cortex Network provides top-down regulation of limbic circuits during emotion regulation, illustrating how "higher" networks can modulate "lower" ones.

Disruptions in network integration characterize many neuropsychiatric disorders. Schizophrenia features reduced coordination between networks supporting cognition and perception, potentially contributing to hallucinations and thought disorder. Autism spectrum disorders show altered patterns of network segregation and integration, which may underlie both enhanced perceptual

Brain networks, including the corticofugal, default mode, dorsal attention, frontoparietal, hippocampal, limbic, salience, somatomotor, ventral attention, visual, and ventral striatum networks, collaborate to regulate motor control, sensory processing, attention, memory, emotional regulation, decision-making, and reward processing.

Each network interacts intricately with others to sustain cognitive, emotional, and behavioral functions. Disruptions in these networks are associated with different neurological and psychiatric disorders.

iv. Types and Subtypes of Brain Cells

The human brain contains approximately 170 billion cells of remarkable diversity, each playing specific roles in the complex symphony of neural function. This taxonomy of primary categories and subcategories of brain cells is integral for understanding cognitive functioning, neural regulation, and the research of neuropsychiatric and neurodegenerative diseases.

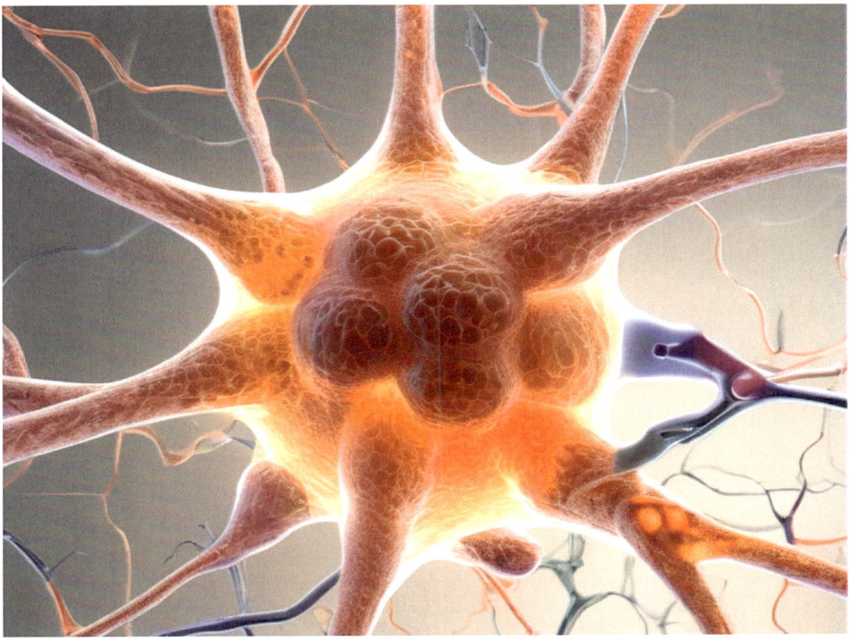

Neuron. Rendering.

With an understanding of these cell types and subtypes, we can investigate the complex workings of the brain. The dysfunction of any of these cellular populations can result in a variety of illnesses, emphasizing their importance in sustaining cognitive and neurological health.

Evolutionary Development of Brain Cell Types

The diversity of brain cells represents a remarkable achievement of evolutionary development. The most fundamental cell types—neurons and basic glial cells—appeared early in the evolution of nervous systems, with recognizable forms present in primitive invertebrates. As more complex organisms evolved, increasing specialization of cell types allowed for more sophisticated neural processing and behavior.

The emergence of oligodendrocytes and myelination represents a critical evolutionary advance in vertebrates, enabling faster signal transmission and more complex neural processing. Specialized interneuron subtypes expanded dramatically in mammals, particularly primates, allowing for more sophisticated regulation of neural circuits. The human brain features not only a greater number but a greater diversity of interneurons than other species, potentially contributing to our unique cognitive capabilities.

Studies of brain development across species have revealed that the longer developmental timeline in humans allows for more extensive expansion and specialization of cell populations. This extended neurodevelopmental period may explain the extraordinary diversity of cell types in the human brain, particularly in regions like the cerebral cortex that underwent significant evolutionary expansion.

Developmental Trajectories of Brain Cells

Brain cells follow remarkably diverse developmental paths from early embryonic life through adulthood. Neural stem cells in the developing brain first generate neurons through neurogenesis, following an inside-out pattern that creates the layered structure of the cortex. Different neuronal subtypes are born at specific developmental windows, with deep layer neurons generated before upper layer neurons.

After the neurogenic period, the same neural stem cells switch to producing glial cells—first astrocytes, then oligodendrocytes. This gliogenic switch is regulated by complex epigenetic mechanisms that change how neural stem cells respond to developmental signals. Microglia follow a unique developmental trajectory, originating from yolk sac progenitors that migrate into the brain during early development.

While most neurogenesis is complete before birth in humans, certain regions—particularly the hippocampus and subventricular zone—retain neurogenic capabilities into adulthood, though at greatly reduced rates. This adult neurogenesis may support ongoing learning and memory functions, with significant implications for cognitive resilience and recovery from injury.

Neurons

Neurons are the basic computational units of the brain and nervous system, responsible for absorbing sensory input, analyzing information, and delivering signals to other neurons, muscles, or glands. They can be classified into several categories according to their function, location, and type of neurotransmitter.

Structural Diversity: Neurons exhibit remarkable structural diversity adapted to specific functions. The human brain contains neurons with dendritic trees of vastly different complexity—from relatively simple structures in some interneurons to the elaborate branching patterns of cortical pyramidal cells that can receive inputs from thousands of other neurons. This structural diversity enables neurons to perform different types of information processing, from simple relay functions to complex integration of diverse inputs.

Molecular Signatures: Modern single-cell transcriptomics has revealed that neurons previously grouped together based on morphology often comprise multiple molecularly distinct subtypes with different gene expression profiles. These molecular signatures influence everything from ion channel expression to neurotransmitter receptor profiles, creating functional diversity even among structurally similar neurons.

Pyramidal Cells

Pyramidal neurons are excitatory neurons located in the cerebral cortex and hippocampus. They are distinguished by their pyramid-shaped cell body and lengthy dendrites. These cells represent approximately 70-80% of all neurons in the cerebral cortex and form the primary output pathways from cortical areas.

Evolutionary Significance: Pyramidal neurons show remarkable evolutionary expansion in primates and humans. Human pyramidal cells have more complex dendritic branching patterns, allowing them to integrate inputs from more diverse sources than those of other species. This expanded integrative capacity may support the more sophisticated cognitive functions characteristic of humans.

Development and Plasticity: Pyramidal neuron development follows a protracted timeline, with dendritic elaboration and spine formation continuing well into early childhood. These neurons remain highly plastic throughout life, with their dendritic spines forming and retracting in response to experience. This structural plasticity provides a physical substrate for learning and memory, with new connections forming as new information is acquired.

Corticofugal neurons, a subtype of pyramidal neurons, transmit information from the cortex to subcortical areas, important in motor control and sensory processing. They form the primary output pathway from the cortex to the thalamus, brainstem, and spinal cord.

Intratelencephalic neurons are cortical neurons that project within the telencephalon (cerebral hemispheres) and are frequently associated with higher cognitive abilities. These neurons form connections between different cortical areas and between the cortex and striatum, supporting complex information integration.

RNA Neuromodulator Applications: Pyramidal neurons express a wide range of receptor types, making them potential targets for RNA neuromodulators. By targeting specific receptor expression in pyramidal neurons, these therapies could potentially normalize circuit activity in conditions characterized by cortical hyperexcitability, such as epilepsy or certain anxiety disorders.

Interneurons

Interneurons are predominantly inhibitory neurons that regulate the activity of other neurons, serving as the brain's "circuit breakers" to prevent runaway excitation. While making up only about 20% of cortical neurons, their diversity and strategic positioning give them outsized influence on neural circuit function.

Evolutionary Development: The diversity of interneuron subtypes expanded dramatically during mammalian evolution, with primates and humans showing greater variety than other species. This expansion suggests the importance of sophisticated inhibitory control for complex cognition. Particular interneuron subtypes are either unique to or greatly expanded in humans, potentially contributing to our distinctive cognitive capabilities.

Developmental Origins: Cortical interneurons originate primarily from the ganglionic eminences in the ventral telencephalon, migrating tangentially to reach the developing cortex. Different interneuron subtypes are born at different developmental time points and from different progenitor zones, contributing to their functional diversity.

GABAergic interneurons secrete gamma-aminobutyric acid (GABA), which is the main inhibitory neurotransmitter in the brain. They shape neural activity through various mechanisms, including feedforward and feedback inhibition, lateral inhibition, and disinhibition. Their dysfunction is implicated in numerous neuropsychiatric disorders, including schizophrenia, autism, and epilepsy.

Subtypes of GABAergic Interneurons Include:

Parvalbumin-expressing interneurons: These fast-spiking cells provide powerful perisomatic inhibition of pyramidal neurons, controlling their output and synchronizing their activity. They play crucial roles in generating gamma oscillations associated with attention and cognitive processing.

Somatostatin-expressing interneurons: These cells target the distal dendrites of pyramidal neurons, regulating their dendritic integration and calcium dynamics. They are involved in the regulation of cortical oscillations and synaptic plasticity, contributing to learning and memory processes.

Vasoactive intestinal peptide (VIP)-expressing interneurons: These cells primarily inhibit other interneurons, creating disinhibition of pyramidal neurons in a circuit mechanism known as disinhibition. They are particularly responsive to signals from other brain regions, helping coordinate local circuit activity with broader brain states.

Chandelier Cells: These specialized interneurons form distinctive "cartridge" synapses onto the axon initial segments of pyramidal neurons, powerfully controlling their output. Dysfunction of chandelier cells has been implicated in schizophrenia, highlighting their importance in normal cognitive function.

RNA Neuromodulator Potential: The diversity of interneuron subtypes offers opportunities for highly specific targeting with RNA neuromodulators. By enhancing or reducing the expression of specific receptors or ion channels in particular interneuron populations, these therapies could potentially rebalance circuit activity with greater precision than traditional pharmaceuticals that broadly affect GABA transmission.

Granule Cells

Granule cells are small, densely packed neurons found in several brain regions, including the cerebellar cortex, dentate gyrus of the hippocampus, and olfactory bulb. Despite their small size, these neurons play crucial roles in information processing and learning.

Evolutionary Context: The basic organization of granule cell circuits shows remarkable evolutionary conservation, suggesting their fundamental importance in neural processing. However, the human dentate gyrus contains a significantly expanded population of granule cells compared to other species, potentially supporting our enhanced capacity for episodic memory.

Developmental Trajectory: Dentate granule cells are among the few neuronal types that continue to be generated throughout adult life in mammals, including humans. This adult neurogenesis declines with age but can be enhanced by exercise, enriched environments, and certain medications, suggesting potential for intervention to support cognitive health.

Dentate granule cells, a subtype of granule cells, play a crucial role in the creation of new memories and the control of mood. They are located in the dentate gyrus of the hippocampus, where they receive inputs from the entorhinal cortex and project to CA3 pyramidal neurons. These cells perform pattern separation—the ability to distinguish between similar experiences or environments—a function critical for episodic memory formation.

Cerebellar granule cells are the most numerous neurons in the brain, forming a dense layer in the cerebellar cortex. They receive inputs from mossy fibers and project to Purkinje cells, playing essential roles in motor learning and coordination.

Network Integration: Granule cells transform information before passing it to principal neurons in their respective circuits. In the hippocampus, granule cells perform sparse coding—where information is represented by the activity of a small subset of neurons—enhancing memory capacity.

Clinical Relevance: Disruptions in dentate granule cell function are implicated in various conditions, including epilepsy, depression, and age-related cognitive decline. In temporal lobe epilepsy, aberrant granule cell wiring may contribute to seizure generation. In depression, reduced hippocampal neurogenesis may impair pattern separation, contributing to overgeneralization of negative experiences.

RNA Neuromodulator Applications: RNA neuromodulators targeting specific receptor systems in dentate granule cells could potentially enhance pattern separation function in conditions like age-related cognitive decline or post-traumatic stress disorder. By modulating NMDA receptor subunit expression, for example, these therapies might normalize granule cell excitability without broadly affecting hippocampal function.

Purkinje Cells

Purkinje cells are among the largest and most distinctive neurons in the brain, with elaborate dendritic trees that fan out like intricate branches in the cerebellar cortex. These enormous neurons discovered in the cerebellar cortex play a vital role in the regulation and coordination of motor activities and are increasingly recognized for their contributions to cognitive and emotional processing as well.

Evolutionary Significance: Purkinje cells show remarkable evolutionary conservation in their basic morphology and circuit organization, highlighting their fundamental importance in motor control across vertebrates. However, human Purkinje cells display more complex dendritic arborization than those of other mammals, potentially supporting more sophisticated motor and cognitive functions.

Developmental Vulnerability: Purkinje cells are generated early in brain development and are particularly vulnerable to developmental insults, including alcohol exposure, hypoxia, and certain genetic mutations. This vulnerability may explain why cerebellar dysfunction is associated with a wide range of developmental disorders, including autism spectrum disorders.

These neurons receive information from both the cerebellar cortex and other regions of the brain, integrating signals from over 100,000 synaptic inputs. Their inhibitory output is involved in regulating the timing and precision of muscle activation, ensuring that movements are executed smoothly and in a coordinated manner.

Beyond motor control, Purkinje cells contribute to cognitive processes through their connections with prefrontal and parietal cortices. Dysfunction of these cells is implicated not only in motor disorders like ataxia but also in conditions affecting cognition and emotion, including autism, schizophrenia, and mood disorders.

Molecular Specialization: Purkinje cells express a unique complement of ion channels, receptors, and signaling molecules that support their specialized functions. They are the only neurons to express certain calcium binding proteins and the only cerebellar neurons to express the metabotropic glutamate receptor mGluR1, making them potential targets for highly specific interventions.

RNA Neuromodulator Potential: The unique molecular profile of Purkinje cells offers opportunities for highly specific targeting with RNA neuromodulators. By enhancing or reducing the expression of particular receptors or ion channels, these therapies could potentially address cerebellar dysfunction in conditions ranging from spinocerebellar ataxias to autism spectrum disorders.

Glial Cells

Glial cells, once thought to be merely supportive elements, are now recognized as active participants in brain function. They outnumber neurons in the human brain and perform various critical functions, from maintaining homeostasis to participating directly in information processing.

Evolutionary Development: Glial diversity increased dramatically during vertebrate evolution, with mammals possessing the most diverse and specialized glial populations. This evolutionary expansion parallels increases in brain size and complexity, suggesting that sophisticated glial functions were necessary for the development of advanced cognitive capabilities.

Developmental Trajectories: Most gliogenesis occurs postnatally in humans, with different glial populations following distinct developmental timelines. Astrocytes and oligodendrocytes continue to mature through childhood and adolescence, potentially contributing to ongoing refinement of neural circuits during development.

Astrocytes

Astrocytes are star-shaped glial cells that perform numerous functions for normal brain operation, from maintaining the blood-brain barrier to regulating neurotransmitter levels at synapses.

Evolutionary Specialization: Human astrocytes are larger, more complex, and more diverse than those of other mammals. When human astrocytes are transplanted into mouse brains, the mice show enhanced learning capabilities, suggesting these specialized cells may contribute to human cognitive prowess.

Functional Diversity: Modern research has revealed substantial heterogeneity among astrocytes, with different subtypes specialized for different functions. Protoplasmic astrocytes in gray matter have different morphology and gene expression patterns than fibrous astrocytes in white matter, reflecting their different roles in supporting neurons versus axons.

Astrocytes regulate synaptic function by removing neurotransmitters from the synaptic cleft, providing energy substrates to neurons, and releasing gliotransmitters that modulate neuronal activity. A single astrocyte can contact and influence thousands of synapses, allowing for coordination of activity across a local neural circuit.

These cells also play crucial roles in brain development and response to injury. During development, astrocytes guide neuronal migration and promote synapse formation. After injury, they can form glial scars that isolate damaged areas.

Clinical Relevance: Astrocyte dysfunction is implicated in numerous neurological and psychiatric disorders. In Alzheimer's disease, reactive astrocytes may contribute to neuroinflammation and amyloid plaque formation. In major depression, reduced astrocytic support may contribute to observed gray matter atrophy and synapse loss.

RNA Neuromodulator Applications: RNA neuromodulators targeting specific genes in astrocytes could potentially normalize their function in conditions like Alzheimer's disease or major depression. By enhancing the expression of glutamate transporters, for example, these therapies might improve glutamate clearance and reduce excitotoxicity without directly affecting neuronal function.

Oligodendrocytes

Oligodendrocytes are specialized glial cells responsible for producing myelin, the fatty insulating sheath that wraps around axons and enables rapid signal transmission in the brain and spinal cord.

Developmental Timeline: Oligodendrocyte precursor cells (OPCs) are generated during late embryonic and early postnatal development and continue to divide and generate new oligodendrocytes throughout life. This ongoing myelination supports new learning and may contribute to cognitive resilience during aging.

Functional Plasticity: Contrary to earlier beliefs, myelin formation is not a fixed developmental process but continues to be dynamically regulated throughout life. Experience-dependent myelination has been observed in response to learning new skills, suggesting that oligodendrocytes actively participate in neural plasticity.

Each oligodendrocyte can myelinate multiple axonal segments, with a single cell capable of wrapping up to 50 different axons. This extensive influence allows a relatively small number of oligodendrocytes to have widespread effects on neural transmission.

Beyond simply insulating axons, oligodendrocytes provide metabolic support to the axons they ensheath and may influence the placement and function of ion channels along the axon. They also participate in adaptive myelination—activity-dependent changes in myelin thickness or coverage that can fine-tune circuit function.

Clinical Significance: Oligodendrocyte dysfunction is central to demyelinating diseases like multiple sclerosis, where autoimmune attacks on myelin disrupt neural transmission. More subtle abnormalities in myelination have been implicated in various psychiatric disorders, including schizophrenia and major depression, where disrupted connectivity between brain regions may contribute to symptoms.

RNA Neuromodulator Potential: RNA neuromodulators targeting genes involved in myelin production or maintenance could potentially enhance remyelination in conditions like multiple sclerosis. By modulating the expression of specific factors in oligodendrocyte precursor cells, these therapies might promote their differentiation into mature, myelinating oligodendrocytes in demyelinated regions.

Microglia

Microglia are the resident immune cells of the central nervous system (CNS), comprising approximately 10% of all brain cells. Unlike other brain cells, microglia originate from yolk sac progenitors that migrate into the developing brain, giving them a unique developmental lineage.

Evolutionary Context: Microglial-like cells are present in all vertebrates, reflecting the fundamental need for immune surveillance in the nervous system. However, human microglia show unique gene expression patterns compared to those of other species, suggesting specialized functions that may relate to our extended lifespan and complex cognitive capabilities.

Developmental Roles: During brain development, microglia actively participate in circuit refinement by pruning excess synapses. They preferentially eliminate less active synapses, following the principle of "use it or lose it" that helps sculpt precise neural circuits based on experience. This developmental pruning is particularly active during adolescence in regions like the prefrontal cortex.

Beyond their traditional role in immune defense, microglia continuously survey the brain parenchyma, extending and retracting processes to monitor their environment. This surveillance allows them to quickly respond to injury or infection, but also enables them to monitor and modulate normal synaptic function.

Microglia exist in various activation states, from "resting" (still active in surveillance) to various "activated" phenotypes. While traditionally categorized as pro-inflammatory (M1) or anti-inflammatory (M2), current research suggests a much more complex spectrum of activation states tailored to specific contexts and signals.

Clinical Relevance: Microglial dysfunction is implicated in numerous neurological and psychiatric disorders. In neurodegenerative diseases like Alzheimer's, aberrant microglial activation may contribute to chronic neuroinflammation and accelerated neuronal death. In autism spectrum disorders, disrupted developmental synaptic pruning by microglia may lead to altered circuit connectivity.

RNA Neuromodulator Applications: RNA neuromodulators targeting specific genes in microglia could potentially modulate their activation state in conditions characterized by neuroinflammation. By reducing the expression of pro-inflammatory factors or enhancing anti-inflammatory pathways, these therapies might shift microglial phenotypes toward more neuroprotective states without broadly suppressing immune function.

Ependymal Cells

Ependymal cells are ciliated epithelial cells that line the ventricles of the brain and the central canal of the spinal cord. They form a barrier between the cerebrospinal fluid (CSF) and brain parenchyma, regulating the flow and composition of CSF.

Developmental Origin: Ependymal cells derive from radial glial cells, which serve as neural stem cells during early brain development. Their transition from radial glia to ependymal cells occurs primarily in the late embryonic and early postnatal periods, coinciding with the expansion of the ventricular system.

Functional Specialization: The beating of ependymal cilia creates currents in the CSF that help distribute signaling molecules and clear metabolic waste. Disruptions in ciliary function can lead to hydrocephalus—excessive accumulation of CSF in the ventricles—highlighting the importance of this seemingly simple function.

Beyond their barrier and transport functions, ependymal cells play roles in regulating neural stem cell function in adjacent regions. In some locations, especially the subventricular zone, they help maintain the neural stem cell niche through the secretion of various factors.

Specialized ependymal cells in structures like the choroid plexus actively produce CSF, which not only provides mechanical protection for the brain but also delivers nutrients and removes waste products. The composition of CSF is tightly regulated by these cells, creating an optimal environment for neural function.

Clinical Significance: Dysfunction of ependymal cells is implicated in conditions like hydrocephalus and certain neurodevelopmental disorders. Loss of ciliary function can disrupt CSF flow, while abnormalities in choroid plexus function may alter CSF composition, potentially affecting brain development and function.

RNA Neuromodulator Potential: While ependymal cells themselves may not be primary targets for RNA neuromodulators in most conditions, understanding their role in maintaining the neural stem cell niche could inform therapies aimed at enhancing neurogenesis or repair after injury. Modulating specific signaling pathways in these cells might create more favorable conditions for endogenous neural stem cells to proliferate and differentiate.

Additional Specialized Cells

The brain contains several specialized cell populations that don't fit neatly into the traditional neuron/glia dichotomy but perform crucial functions in neural development and function.

Neural Stem Cells and Progenitors

Neural stem cells persist in specific niches in the adult brain, particularly the subventricular zone lining the lateral ventricles and the subgranular zone of the hippocampal dentate gyrus. They retain the ability to generate new neurons and glial cells throughout life.

The cells of the *Rostral Migratory Stream* are involved in the migration of neuronal progenitors from the subventricular zone to the olfactory bulb, playing a role in the ongoing neurogenesis and olfactory function. This remarkable migration occurs along a specialized pathway where chains of new neurons travel through tunnels formed by astrocytes to reach their destination.

Developmental Significance: Neural stem cells undergo a series of fate restrictions during development, transitioning from multipotent progenitors capable of generating all neural cell types to more restricted progenitors with limited potential. This progressive specification ensures the generation of appropriate cell types in the correct proportions and locations.

Environmental Regulation: Neural stem cell activity is regulated by factors, including exercise, stress, and inflammation. Physical activity enhances hippocampal neurogenesis, contributing to exercise's positive effects on cognition, while chronic stress suppresses it, possibly contributing to stress-related cognitive and mood disorders.

Clinical Relevance: Dysregulation of neural stem cells is implicated in various conditions, from neurodevelopmental disorders like autism to age-related cognitive decline. The limited regenerative capacity of these cells presents both a challenge and an opportunity for therapeutic interventions aimed at enhancing brain repair.

RNA Neuromodulator Applications: RNA neuromodulators targeting specific signaling pathways in neural stem cells could potentially enhance their proliferation or direct their differentiation toward specific cell types. By modulating factors that inhibit neurogenesis, for example, these therapies might promote the generation of new neurons in conditions like traumatic brain injury.

Tanycytes

Tanycytes are specialized ependymal cells lining portions of the third ventricle, particularly adjacent to the hypothalamus. They have long processes that extend into the hypothalamic parenchyma, allowing them to sample both ventricular CSF and brain tissue.

Functional Significance: Tanycytes serve as sensors for various circulating factors, including nutrients and hormones, and relay this information to hypothalamic circuits involved in energy balance and metabolism. They may also transport molecules from the CSF into the brain, bypassing the blood-brain barrier.

Some populations of tanycytes retain neural stem cell properties in adulthood and can generate new neurons in the hypothalamus. This adult hypothalamic neurogenesis may play roles in energy balance and metabolic regulation, with potential implications for conditions like obesity.

Clinical Relevance: Dysfunction of tanycytes is implicated in disorders of energy balance and metabolism, including obesity and diabetes. Understanding their role as mediators between peripheral signals and central regulatory circuits may offer new approaches to treating these increasingly prevalent conditions.

Cells With Functional Significance Across Networks

Different cell types interact within and across brain networks to support various aspects of cognition and behavior. Understanding these interactions is essential for developing targeted interventions for neurological and psychiatric disorders.

Pyramidal cells have a role in cognitive processes such as decision-making, planning, and remembering. Their function is to control the equilibrium between excitatory and inhibitory impulses, which is essential for preserving the stability of brain circuits. The long-range projections of pyramidal neurons form the primary pathways for communication between different brain regions, enabling the coordination of activity across distributed networks.

Astrocytes, oligodendrocytes, and microglia are types of cells found in the central nervous system and play crucial functions in facilitating neuronal activity, preserving the extracellular space, generating myelin, and engaging in immunological responses. Their coordinated activity with neurons ensures optimal brain function across multiple time scales, from millisecond-level synaptic transmission to long-term circuit plasticity and remodeling.

Cell-Type Specific Targeting with RNA Neuromodulators

The remarkable diversity of brain cell types offers both challenges and opportunities for therapeutic interventions. Traditional pharmaceuticals often affect multiple cell types simultaneously, leading to unwanted side effects. RNA neuromodulators, in contrast, can potentially target specific cell populations through several mechanisms:

1. *Cell-Type Specific Promoters:* RNA delivery vectors can incorporate promoters that are only active in specific cell types, ensuring that the RNA is only expressed in target populations. For example,

using the CaMKII promoter would restrict expression primarily to excitatory neurons.

2. *Receptor/Channel Targeting:* Even within broadly distributed cell types, RNA neuromodulators can target receptors or channels with restricted expression patterns. By targeting a receptor subtype only expressed in certain brain regions or cell populations, these therapies can achieve functional specificity.

3. *Anatomical Targeting:* The intranasal delivery route combined with engineered viral vectors can achieve preferential targeting of specific brain regions, allowing for spatial specificity in RNA delivery.

This precision targeting represents a significant advance over traditional psychopharmacology, potentially allowing for the correction of specific circuit dysfunctions without broadly affecting brain function. As our understanding of the molecular signatures of different brain cell types continues to expand, opportunities for increasingly precise interventions will emerge, potentially transforming our approach to treating neurological and psychiatric disorders.

v. Types and Functions of Neuronal Receptors

Neuronal receptors represent the fundamental interface between cells and their environment, serving as sophisticated molecular machines that convert chemical signals into cellular responses.

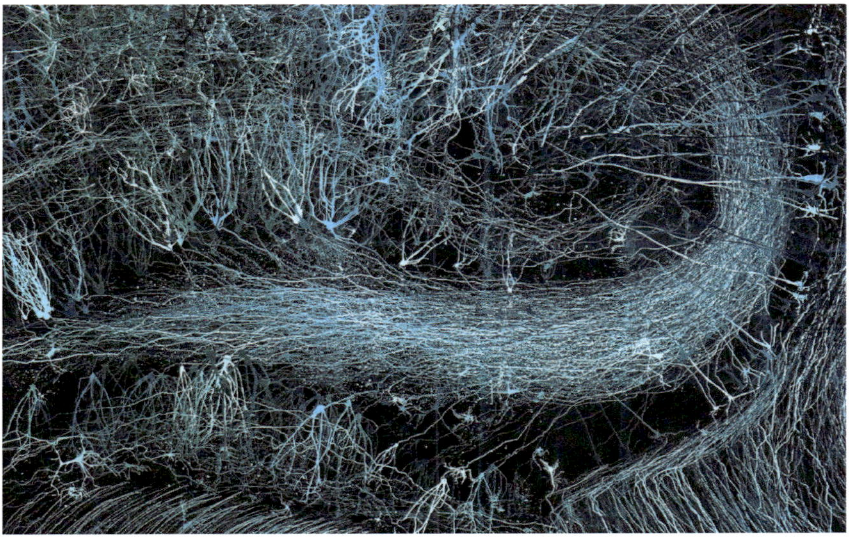

Dentate Gyrus, Greg Dunn

These remarkable structures not only facilitate communication between neurons but also determine how our brains process information, regulate mood, form memories, and respond to both internal and external stimuli. Understanding these receptors provides crucial insights into normal brain function and offers targets for RNA neuromodulators and other therapeutic interventions.

5-HT2a Receptors

5-HT2a receptors are a subtype of serotonin receptors expressed on specific neuronal populations throughout the cortex, with particularly high density in layer V pyramidal neurons. These receptors play a crucial role in modulating mood, cognition, and perception by coupling to Gq/G11 proteins and activating phospholipase C signaling cascades.

Evolutionary Significance: The serotonergic system, including 5-HT2a receptors, emerged early in vertebrate evolution, with highly conserved structures across species. This conservation suggests their fundamental importance in neural function. The 5-HT2a receptor subtype appears to have evolved specialized roles in higher cognitive functions as brains became more complex, helping organisms process and integrate complex sensory and cognitive information.

Structural Characteristics: The 5-HT2a receptor consists of seven transmembrane domains with an extracellular N-terminus and intracellular C-terminus. The binding pocket for serotonin is formed by residues from multiple transmembrane segments, creating a specific three-dimensional environment that determines ligand selectivity.

The effects of psychedelic drugs, such as LSD and psilocybin, are closely associated with their ability to bind to and activate 5-HT2a receptors, particularly in cortical areas. These compounds act as agonists at the receptor, producing altered states of consciousness characterized by perceptual changes, emotional intensification, and cognitive flexibility.

Clinical Significance: Beyond their role in psychedelic effects, 5-HT2a receptors play a major role in the pathophysiology of several mental disorders. In schizophrenia, abnormal 5-HT2a receptor function may contribute to both positive and negative symptoms. Many atypical antipsychotics act as 5-HT2a antagonists, blocking these receptors as part of their therapeutic mechanism.

In depression and anxiety disorders, 5-HT2a receptor dysregulation is increasingly recognized as a contributor to symptom development. Recent research has demonstrated that the rapid antidepressant effects of psychedelic compounds may relate to their action on 5-HT2a receptors, triggering neuroplastic changes that can reset dysfunctional circuit activity.

Receptor Plasticity: 5-HT2a receptors exhibit remarkable plasticity, with their expression levels and sensitivity changing in response to stress, medications, and environmental factors. Chronic stress can alter 5-HT2a receptor density, potentially contributing to stress-related disorders. Conversely, some antidepressant treatments downregulate these receptors over time, suggesting complex adaptive mechanisms.

RNA neuromodulators targeting 5-HT2a receptors offer a novel approach to treating conditions like treatment-resistant depression and anxiety by precisely regulating receptor expression in specific brain regions, potentially avoiding the side effects associated with systemic pharmacological interventions.

GABA Receptors

GABA (gamma-aminobutyric acid) receptors play a crucial role in inhibitory neurotransmission throughout the central nervous system. The two main types, GABA_A and GABA_B receptors, mediate different aspects of inhibitory signaling.

Evolutionary Development: GABA emerged as a signaling molecule in prokaryotes billions of years ago, long before the development of complex nervous systems. As multicellular organisms evolved, GABA was co-opted for intercellular communication, eventually becoming the primary inhibitory neurotransmitter in vertebrate brains. This ancient heritage explains why GABA receptors are found across virtually all brain regions and neuronal types, providing essential inhibitory balance to excitatory signaling.

GABA_A receptors are ligand-gated ion channels that facilitate rapid inhibition by enabling the influx of chloride ions into the neuron. This results in hyperpolarization, decreasing the probability of action potential generation. These pentameric receptors are composed of different subunit combinations (α, β, γ, δ, ε, θ, π, and ρ), with each combination conferring distinct pharmacological and kinetic properties.

GABA_B receptors, in contrast, are G-protein-coupled receptors that produce slower and longer-lasting inhibitory effects through second messenger systems. They operate primarily by inhibiting calcium channels and activating potassium channels, resulting in reduced neurotransmitter release and neuronal hyperpolarization.

Clinical Relevance: Benzodiazepines and barbiturates primarily target GABA_A receptors, enhancing the inhibitory effect of endogenous GABA. These medications play crucial roles in controlling anxiety, inducing sedation, and promoting muscle relaxation. Dysfunction in GABAergic signaling has been implicated in numerous neurological and psychiatric conditions, including epilepsy, anxiety disorders, insomnia, and schizophrenia.

Receptor Integration: GABA receptors don't function in isolation but operate within complex neural networks where they interact with excitatory systems to maintain proper excitation-inhibition balance. For example, GABA interneurons often form feed-forward inhibitory circuits with glutamatergic neurons, creating precise temporal windows for information processing and preventing runaway excitation.

RNA neuromodulators targeting specific GABA receptor subunits could potentially offer more precise treatment approaches for conditions like epilepsy or anxiety disorders, where global enhancement of GABAergic transmission via traditional pharmaceuticals often produces unwanted sedation or cognitive impairment.

Glutamate Receptors

Glutamate receptors are essential for fast excitatory neurotransmission in the central nervous system. Their involvement facilitates synaptic plasticity, which serves as the foundation for learning and memory. The dysregulation of these receptors has been associated with several neurological disorders, including Alzheimer's disease, epilepsy, and amyotrophic lateral sclerosis (ALS).

Evolutionary Context: Glutamate signaling evolved early in the development of nervous systems, with glutamate receptors appearing in primitive invertebrates. The diversification of glutamate receptor subtypes paralleled the evolution of more complex brains, allowing for increasingly sophisticated information processing and storage. This evolutionary trajectory highlights the fundamental importance of glutamatergic transmission in cognitive functions.

Glutamate receptors may be classified into three main categories of ionotropic receptors: NMDA receptors, AMPA receptors, and kainate receptors, plus metabotropic glutamate receptors (mGluRs). Each of these receptors has a unique role in synaptic transmission and plasticity.

NMDA Receptors

NMDA (N-methyl-D-aspartate) receptors are a subtype of glutamate receptors that exert a substantial impact on both synaptic plasticity and memory formation. Their unique properties make them molecular coincidence detectors in neural circuits.

Molecular Structure: NMDA receptors are tetrameric complexes typically composed of two GluN1 subunits combined with two GluN2 subunits (which come in four subtypes: GluN2A-D). This structural diversity allows for regional specialization of NMDA receptor function throughout the brain.

A distinguishing feature of NMDA receptors is their dual activation requirement: they need both the binding of glutamate and glycine (as co-agonists) and the depolarization of the cell membrane to remove a magnesium ion block from the channel pore. When activated, they allow calcium ions to enter the cell, triggering various intracellular signaling cascades.

This unique activation mechanism gives NMDA receptors a crucial function in long-term potentiation (LTP) and long-term depression

(LTD), which are vital for learning and memory. By detecting the coincidence of presynaptic activity (glutamate release) and postsynaptic activity (membrane depolarization), NMDA receptors can strengthen or weaken specific synaptic connections based on experience.

Clinical Implications: Overstimulation of NMDA receptors can lead to excitotoxicity, a process implicated in neurodegenerative conditions such as Alzheimer's disease, stroke, and traumatic brain injury. Conversely, NMDA receptor hypofunction has been linked to schizophrenia, with NMDA receptor antagonists like ketamine producing schizophrenia-like symptoms in healthy individuals.

Therapeutic Targeting: NMDA receptors represent important targets for various neurological conditions. Memantine, an NMDA receptor antagonist, is used to treat moderate-to-severe Alzheimer's disease. More recently, ketamine's action as an NMDA receptor antagonist has shown rapid antidepressant effects, revolutionizing approaches to treatment-resistant depression.

RNA neuromodulators targeting specific NMDA receptor subunits could potentially offer more precise interventions for conditions characterized by glutamatergic dysfunction, avoiding the broad effects of traditional NMDA receptor drugs.

AMPA Receptors

AMPA (α-amino-3-hydroxy-5-methyl-4-isoxazolepropionic acid) receptors are ionotropic glutamate receptors that mediate fast excitatory synaptic transmission in the brain.

Structure and Function: AMPA receptors are tetrameric complexes composed of combinations of four subunits (GluA1-4). Different subunit compositions confer distinct functional properties, including ion permeability and synaptic integration. Most AMPA receptors are permeable to sodium and potassium ions, but some can also allow calcium entry, depending on their subunit composition.

AMPA receptors play a crucial role in transmitting most of the rapid excitatory signals in the brain. Additionally, they are actively regulated during synaptic plasticity, with their trafficking to and from the synapse representing a major mechanism for strengthening or weakening synaptic connections.

Plasticity Mechanisms: During LTP, AMPA receptors are rapidly inserted into the postsynaptic membrane, increasing the synapse's sensitivity to glutamate. Conversely, during LTD, AMPA receptors are removed from the synapse, decreasing its strength. These dynamic changes in AMPA receptor localization and function underlie many forms of learning and memory.

Therapeutic Potential: Compounds that enhance AMPA receptor function (AMPA receptor positive allosteric modulators or "AMPA-kines") have shown promise for treating cognitive deficits in various conditions, including Alzheimer's disease, ADHD, and schizophrenia. RNA neuromodulators targeting specific AMPA receptor subunits could potentially offer more precise approaches to enhancing cognitive function in these conditions.

Dopamine Receptors: D1-D5

Dopamine receptors play a crucial role in a wide range of activities, such as mood regulation, motor control, and motivation. All dopamine receptors are G-protein-coupled receptors, but they are divided into two main families based on their structural and pharmacological properties.

Evolutionary Significance: The dopaminergic system evolved as a critical component of reward processing and motor control in vertebrates. Dopamine signaling is evolutionarily ancient, with similar systems found in organisms as distant as insects. This conservation highlights dopamine's fundamental role in driving approach behaviors toward beneficial stimuli and avoiding harmful ones—a basic function necessary for survival across animal species.

The D1-like family (including D1 and D5 receptors) couples to Gs proteins, activating adenylyl cyclase and increasing cyclic AMP (cAMP) production. These receptors are expressed in various brain regions, including the striatum, prefrontal cortex, and hippocampus, where they generally produce excitatory effects on neurons.

The D2-like family (including D2, D3, and D4 receptors) couples to Gi/Go proteins, inhibiting adenylyl cyclase and decreasing cAMP production. These receptors are expressed in the striatum, substantia nigra, ventral tegmental area, and cortex, where they often produce inhibitory effects on neurons.

Circuit Integration: Dopamine receptors work in concert with other neurotransmitter systems to modulate neural circuits. For example, in the striatum, D1 and D2 receptors are preferentially expressed on different populations of medium spiny neurons, giving rise to the direct and indirect pathways of motor control, respectively. This segregation allows dopamine to bidirectionally modulate movement: activating the direct pathway (via D1 receptors) facilitates movement, while activating the indirect pathway (via D2 receptors) inhibits movement.

Clinical Significance: Disruptions of dopamine signaling have been associated with several neurodegenerative and psychiatric disorders. In Parkinson's disease, the degeneration of dopaminergic neurons leads to motor symptoms that are treated with dopamine replacement strategies or dopamine receptor agonists. In schizophrenia, excessive dopamine signaling in certain brain regions may contribute to positive symptoms, which are treated with D2 receptor antagonists.

Dopamine's role in reward and reinforcement makes its receptors central to addiction processes. Drugs of abuse often increase dopamine release in the nucleus accumbens, creating reinforcement that drives continued drug-seeking behavior. Understanding how specific dopamine receptor subtypes contribute to addiction may lead to more targeted therapies. RNA neuromodulators targeting specific dopamine receptor subtypes in defined brain regions potentially offer more precise treatments for conditions like Parkinson's disease or schizophrenia avoiding side effects associated with systemic dopaminergic medications.

Cholinergic Receptors

Acetylcholine was the first neurotransmitter to be discovered, and its receptors play diverse roles in both the central and peripheral nervous systems. Cholinergic receptors are divided into two main classes: nicotinic and muscarinic receptors.

Evolutionary Development: Cholinergic signaling is one of the most ancient neurotransmitter systems, present even in primitive organisms lacking a central nervous system. This ancient heritage explains why acetylcholine serves such diverse functions, from neuromuscular transmission to cognitive processing. The divergence into nicotinic (ionotropic) and muscarinic (metabotropic) receptor types occurred early in evolution, allowing for both rapid and sustained signaling through the same neurotransmitter.

Nicotinic Acetylcholine Receptors

Nicotinic acetylcholine receptors (nAChRs) are ligand-gated ion channels found in both the central nervous system and the peripheral nervous system. They mediate fast synaptic transmission at the neuromuscular junction and in various brain regions.

Structure and Function: nAChRs are pentameric channels composed of combinations of alpha (α1-α10) and beta (β1-β4) subunits. Different subunit compositions confer distinct functional properties and regional distributions. When activated by acetylcholine or agonists like nicotine, these receptors open, allowing the influx of sodium and calcium ions and the efflux of potassium ions, generally resulting in neuronal excitation.

nAChRs play crucial roles in several cognitive processes, including attention, learning, and memory. They modulate the release of various neurotransmitters, including dopamine, serotonin, and glutamate, influencing circuit activity throughout the brain.

Clinical Relevance: The addictive properties of nicotine can be attributed to its interaction with nAChRs, particularly those containing α4 and β2 subunits, which are highly expressed in the brain's reward pathways. Activation of these receptors ultimately increases dopamine release in the nucleus accumbens, reinforcing nicotine-seeking behavior.

An association has been demonstrated between neurodegenerative diseases, such as Alzheimer's, and the disruption of nicotinic receptors. Alzheimer's patients show significant losses of nAChRs, particularly those containing the α7 subunit, which correlate with cognitive decline. Consequently, nAChR agonists are being investigated as potential treatments for cognitive impairment in Alzheimer's disease.

RNA neuromodulators targeting specific nAChR subunits could potentially offer more precise approaches to treating conditions like nicotine addiction or cognitive decline, without the side effects associated with less selective cholinergic drugs.

Muscarinic Acetylcholine Receptors

Muscarinic acetylcholine receptors (mAChRs) are G-protein-coupled receptors that mediate slower, more sustained responses to acetylcholine compared to nicotinic receptors.

Structure and Function: There exist five distinct subtypes of muscarinic receptors, labeled as M1 through M5. M1, M3, and M5 receptors couple primarily to Gq/11 proteins, activating phospholipase C and increasing intracellular calcium, while M2 and M4 receptors couple to Gi/o proteins, inhibiting adenylyl cyclase and decreasing cAMP production.

These receptors are involved in a range of activities, including cognitive processing, motor control, and regulation of the autonomic nervous system. In the brain, muscarinic receptors modulate neuronal excitability, synaptic plasticity, and the release of various neurotransmitters.

Clinical Applications: Patients with conditions such as schizophrenia and Parkinson's disease often receive treatment through the administration of drugs that target muscarinic receptors. In Parkinson's disease, anticholinergic drugs that block muscarinic receptors can help reduce tremor and rigidity by rebalancing cholinergic and dopaminergic signaling in the basal ganglia.

Conversely, cholinesterase inhibitors, which increase acetylcholine levels and thus enhance muscarinic receptor activation, are used to treat cognitive symptoms in Alzheimer's disease. More selective muscarinic agonists, particularly those targeting the M1 receptor, are being investigated as potential cognitive enhancers with fewer peripheral side effects.

RNA neuromodulators targeting specific muscarinic receptor subtypes could potentially offer more precise therapeutic approaches for conditions characterized by cholinergic dysfunction, avoiding the broad effects of current cholinergic medications.

Opioid Receptors

Opioid receptors are G-protein-coupled receptors that respond to both endogenous opioids (like endorphins, enkephalins, and dynorphins) and exogenous opioids (like morphine and other opiate drugs). They play critical roles in pain modulation, reward processing, and stress responses.

Evolutionary Context: The opioid system evolved primarily as a pain regulation mechanism, allowing organisms to modulate their response to potentially damaging stimuli. This system became increasingly sophisticated in mammals, where it also integrated with reward processing to reinforce beneficial behaviors. The presence of multiple receptor subtypes with distinct functions represents an evolutionary adaptation that allows for nuanced control over various physiological processes.

The classification of opioid receptors can be divided into three main categories: mu (μ), delta (δ), and kappa (κ). Each receptor type has a distinct distribution, pharmacology, and functional role.

Mu opioid receptors are primarily responsible for mediating analgesia, euphoria, and respiratory depression. They are widely distributed in brain regions involved in pain processing, reward, and respiratory control. The pain-relieving and mood-enhancing actions of clinical opioids are primarily associated with mu receptors. However, these same receptors are involved in the development of tolerance and dependence on opioids.

Delta opioid receptors are involved in analgesia, antidepressant effects, and neuroprotection. They are found in brain regions associated with pain, emotion, and cognitive function.

Kappa opioid receptors mediate analgesia, dysphoria, and psychotomimetic effects. They are expressed in brain regions involved in pain, mood, and perception. Activation of kappa receptors typically produces effects opposite to mu receptor activation, including dysphoria rather than euphoria.

Clinical Significance: Opioid receptors are the targets of both the most effective analgesics and some of the most addictive substances. Understanding the specific roles of different opioid receptor subtypes has led to efforts to develop more selective drugs that maintain analgesic efficacy while reducing unwanted effects like respiratory depression and addiction potential.

RNA neuromodulators targeting specific opioid receptor subtypes in defined brain regions could potentially revolutionize pain management by providing precise control over pain pathways without affecting reward or respiratory circuits, potentially avoiding the addiction and safety issues associated with traditional opioids.

Cannabinoid Receptors: CB1 and CB2

Cannabinoid receptors are a component of the endocannabinoid system, which regulates numerous physiological processes, including pain, mood, appetite, and memory. This system consists of the cannabinoid receptors, their endogenous ligands (endocannabinoids), and the enzymes responsible for endocannabinoid synthesis and degradation.

Evolutionary Development: The endocannabinoid system is evolutionarily ancient, with cannabinoid receptors identified in species ranging from simple invertebrates to humans. This conservation suggests fundamental roles in organismal function. Interestingly, the system evolved not to interact with cannabis plants but to respond to endogenous cannabinoids produced within the body. The plant simply evolved compounds that mimic these endogenous signals.

Structure and Function: CB1 receptors are primarily located in the brain, particularly in regions involved in cognition, memory, anxiety, pain perception, and motor coordination. They are among the most abundant G-protein-coupled receptors in the brain. CB1 receptors typically couple to Gi/o proteins, inhibiting adenylyl cyclase and reducing cAMP production. They also modulate ion channels, typically inhibiting calcium channels and activating potassium channels.

A unique feature of CB1 receptor signaling is its predominantly retrograde direction: endocannabinoids are typically released from postsynaptic neurons and act on CB1 receptors located on presynaptic terminals, where they inhibit neurotransmitter release. This mechanism allows for activity-dependent modulation of synaptic strength.

CB2 receptors are widely distributed throughout the immune system and have been associated with the regulation of pain and inflammation. They are also expressed in some central nervous system cells, particularly microglia, where they modulate neuroinflammatory responses. Like CB1 receptors, CB2 receptors typically couple to Gi/o proteins, inhibiting adenylyl cyclase.

Signaling Integration: The endocannabinoid system doesn't operate in isolation but interacts with other neurotransmitter systems. For example, endocannabinoids can modulate dopamine release in the nucleus accumbens, influencing reward processing. Similarly, interactions between the endocannabinoid and opioid systems contribute to pain modulation.

Clinical Relevance: Both endogenous cannabinoids, such as anandamide and 2-arachidonoylglycerol (2-AG), and exogenous cannabinoids, such as THC derived from cannabis, interact with cannabinoid receptors to produce their effects.

The therapeutic potential of cannabinoid receptor modulation is being explored for various conditions, including chronic pain, epilepsy, neurodegenerative disorders, and psychiatric conditions. Selective modulators of CB1 and CB2 receptors may offer therapeutic benefits without the psychoactive effects associated with non-selective cannabinoid receptor activation.

RNA neuromodulators targeting cannabinoid receptors could potentially offer precise control over specific aspects of the endocannabinoid system, allowing for therapeutic benefits in conditions like chronic pain or epilepsy without the broad effects of cannabis or synthetic cannabinoids.

Adrenergic Receptors

Adrenergic receptors become active when exposed to the neurotransmitters norepinephrine (noradrenaline) and epinephrine (adrenaline). These receptors play crucial roles in the "fight-or-flight" response and in regulating various autonomic functions.

Evolutionary Significance: The adrenergic system evolved as a critical component of the stress response in vertebrates, allowing rapid physiological adaptations to threatening situations. The diversification of adrenergic receptor subtypes allowed for more nuanced control over various organ systems, enabling complex and coordinated responses to environmental challenges.

Adrenergic receptors can be classified into two main types: alpha and beta, each with several subtypes. Alpha receptors are further divided into alpha-1 (α1) and alpha-2 (α2) subtypes, while beta receptors include beta-1 (β1), beta-2 (β2), and beta-3 (β3) subtypes.

Alpha-1 receptors couple to Gq proteins, activating phospholipase C and increasing intracellular calcium. They are expressed in smooth muscle, blood vessels, the heart, and the brain, where they typically produce excitatory effects.

Alpha-2 receptors couple to Gi proteins, inhibiting adenylyl cyclase and decreasing cAMP production. They function as autoreceptors on noradrenergic neurons, where they inhibit norepinephrine release, and are also expressed in various target tissues.

Beta receptors couple to Gs proteins, activating adenylyl cyclase and increasing cAMP production. Beta-1 receptors are predominantly expressed in the heart, beta-2 receptors in the lungs and blood vessels, and beta-3 receptors in adipose tissue.

Functional Roles: Adrenergic receptors regulate various physiological responses during stress, including increasing heart rate, dilating pupils, and mobilizing energy stores. In the brain, adrenergic signaling, particularly through norepinephrine, modulates attention, arousal, and memory consolidation.

Clinical Applications: Adrenergic receptors are targets for numerous medications used to treat cardiovascular conditions, asthma, and other disorders. Beta blockers, which antagonize beta adrenergic receptors, are used to treat hypertension, angina, and certain arrhythmias. Alpha-2 agonists like clonidine are used to treat hypertension and attention deficit hyperactivity disorder (ADHD).

In the central nervous system, drugs targeting adrenergic receptors are used for various conditions. For example, alpha-2 agonists can enhance attention and reduce impulsivity in ADHD, while beta blockers can reduce anxiety symptoms, particularly physical manifestations like tremor and tachycardia.

RNA neuromodulators targeting specific adrenergic receptor subtypes in defined tissues could potentially offer more precise control over adrenergic signaling, allowing for therapeutic benefits without the broad effects of current adrenergic drugs.

Receptor Cross-Talk and Integration

While individual receptor systems have been discussed separately, it's essential to recognize that they don't function in isolation. Instead, neuronal receptors form integrated networks where signaling through one receptor type can modulate the function of others, creating complex and context-dependent responses.

For example, serotonin and glutamate receptors exhibit significant cross-talk in regions like the prefrontal cortex and hippocampus. Activation of 5-HT2a receptors can modulate NMDA receptor function, potentially explaining how psychedelics affect cognitive flexibility and perception.

This receptor cross-talk allows for nuanced control over neural circuit function and may represent an important target for next-generation therapeutics. RNA neuromodulators that target specific receptor subtypes could potentially rebalance these integrated signaling networks in a more precise manner than traditional pharmaceuticals.

Future Directions: RNA Neuromodulators and Receptor-Targeted Therapies

Understanding the diverse array of neuronal receptors provides a foundation for developing increasingly precise therapeutic interventions. RNA neuromodulators offer a promising approach to targeting specific receptor populations in defined brain regions, potentially allowing for more effective treatments with fewer side effects.

By modulating the expression of specific receptor subtypes, RNA neuromodulators could potentially normalize dysregulated neural circuits in various neurological and psychiatric conditions. For example, selectively reducing 5-HT2a receptor expression in hyperactive prefrontal circuits might alleviate anxiety symptoms without affecting serotonergic signaling throughout the brain. Similarly, modulating specific NMDA receptor subunits in the hippocampus could potentially enhance memory without risking excitotoxicity.

The continued advancement of RNA delivery technologies, combined with deepening understanding of receptor function in health and disease, opens exciting possibilities for precision neuromedicine. As these technologies mature, they may transform our approach to treating brain disorders, moving from broad pharmacological interventions to precisely targeted genetic modulation of specific receptor systems in the neural circuits.

vi. Brain Chemistry and Serotonin Biology

The Role of Serotonin

When we consider the complex symphony of brain chemistry that governs our thoughts, emotions, and behaviors, serotonin stands out as one of the most fascinating and consequential conductors. This chapter explores the strategic and scientific rationale behind our focus on the 5-HT2a receptor as a target for RNA therapeutics, beginning with an understanding of serotonin's fundamental role in neural communication.

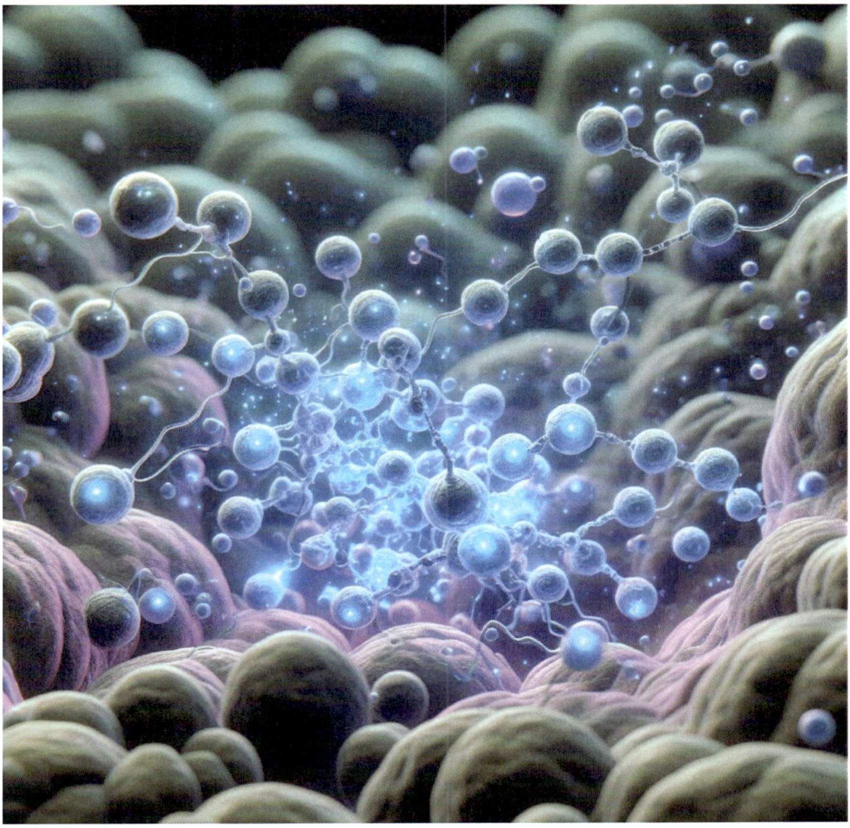

Serotonin Molecules. Rendering.

Neurotransmitters are brain chemicals that facilitate communication between neurons, influencing mood, cognition, and critical physiological functions. Among these, serotonin plays a particularly crucial role in regulating mood, appetite, sleep, and memory. Research has revealed a connection between serotonin imbalances and many mental health disorders, suggesting that this neuromodulator is central to psychological well-being.

I recall visiting a neuroscience laboratory where researchers were mapping serotonin pathways in the brain. "We used to think of neurotransmitters as simple on-off switches," the lead scientist explained, "but we now understand that serotonin functions more like a complex dimmer switch, finely tuning neural circuits throughout the brain." This analogy helps us appreciate why broad-spectrum medications affecting serotonin often produce such varied effects—they're adjusting a sophisticated control system that touches virtually every aspect of our mental and physical experience.

Serotonin receptors, distributed throughout the central and peripheral nervous systems, mediate these effects across a wide range of psychological and physiological processes. Currently, fifteen distinct subtypes of serotonin receptors have been identified in the central nervous system (CNS), each contributing to the intricate control of these processes in unique ways.

Serotonin Receptor Diversity

The multiple functions of serotonin receptors are categorized into seven major families, ranging from 5-HT1 to 5-HT7. Each family has unique subtypes and performs distinctive roles in both the central and peripheral nervous systems. Beyond their effects on mood and cognition, these receptors also influence the operation of the gastrointestinal tract, cardiovascular system, and other physiological processes.

The decrease in serotonin receptor populations caused by SSRI therapies can result in various adverse consequences. These effects frequently stem from receptor downregulation and the accompanying alterations in neurotransmission that occur as the body strives to regain equilibrium. Understanding the specific functions of each receptor subtype helps clarify why conventional treatments often come with a complex profile of side effects. It also illuminates why targeted approaches, such as RNA therapeutics focused on specific receptor subtypes, hold such promise.

5-HT2a Receptors

The 5-HT2a receptors are widely found in the hippocampus, cortex, and subcortical areas. They are crucial in regulating neurotransmission and neuronal excitability, directly affecting cognitive functioning, emotional regulation, and memory formation. Diminished activation of these receptors by serotonin might result in reduced arousal, potentially causing a drop in sexual attraction, increased anxiety, and difficulty sleeping.

Dr. James Chen, a neuropsychiatrist specializing in treatment-resistant mood disorders, describes the 5-HT2a receptor as "a central switchboard for emotional processing." In his practice, he has observed that patients with anxiety often show patterns of 5-HT2a receptor hyperactivity in brain imaging studies. "These receptors appear to be overactive in certain brain regions, essentially amplifying threat signals and dampening the brain's ability to regulate fear responses," he explains. This observation aligns with the scientific rationale for targeting these specific receptors with highly precise RNA therapeutics.

5-HT2B Receptors

The 5-HT2B receptors substantially impact the movement of the gastrointestinal tract and cardiac activities, as they are primarily located in the stomach and heart. The reduction in activity of these receptors can profoundly affect cardiovascular health and gastrointestinal function, resulting in disorders such as pulmonary hypertension, poor gut motility, nausea, vomiting, fewer bowel movements, and decreased bone density—all of which present substantial health risks.

5-HT2C Receptors

The 5-HT2C receptors, found in the choroid plexus and various brain regions, regulate food intake, mood, and anxiety levels. A decrease in the number of these receptors may lead to an enhanced desire to eat and a consequent rise in body weight due to the reduced ability of serotonin to regulate hunger. This modification might also induce metabolic alterations and potentially affect symptoms of anxiety.

Many patients taking SSRIs report weight gain as a troubling side effect. Michael, a 35-year-old software engineer, gained nearly 30 pounds during his first year on an SSRI prescribed for panic disorder. "The medication helped with the panic attacks, but I found myself constantly hungry and craving carbohydrates," he reported.

"It felt like my body's satiety signals were somehow muted." This common side effect illustrates how broad-spectrum serotonergic medications can impact multiple receptor systems simultaneously, producing unwanted effects by modulating receptors like 5-HT2C that weren't the primary therapeutic target.

5-HT3 Receptors

The 5-HT3 receptors, which function as ligand-gated ion channels, are predominantly located in the gastrointestinal tract and specific brain areas linked to feelings of nausea and anxiety. A decline in these receptor populations might diminish their impact on gastrointestinal function, potentially reducing nausea and vomiting.

5-HT4, 5-HT5, 5-HT6, and 5-HT7 Receptors

These specific subtypes of serotonin receptors each have unique involvements in processes such as learning, memory, and gastrointestinal motility. Each contributes to maintaining both neurological and physiological well-being in distinct ways. For example, the 5-HT6 receptor is highly expressed in the striatum and cortex, modulating neuroplasticity. This receptor is crucial for cognitive flexibility and memory formation.

Each of the serotonin receptor subtypes—including 5-HT2a, 5-HT1B/1D, 5-HT2a, 5-HT2B, 5-HT2C, 5-HT3, 5-HT4, 5-HT5, 5-HT6, and 5-HT7—is essential for a particular range of physiological and psychological functions. Changes in their populations or functioning can significantly impact mood, cognition, gastrointestinal health, and general well-being. This complexity explains why conventional psychiatric medications often produce such varied side effect profiles and why precision approaches targeting specific receptor subtypes represent a significant advance in treatment strategy.

Targeting the 5-HT2a Receptor

The 5-HT2a receptor is integral to how the brain processes emotions, stress, and cognition. Its activation alters multiple neuromodulator systems, including dopamine and glutamate, which are crucial for mood regulation and cognitive functions. This makes it a particularly valuable target for therapeutic intervention in conditions involving mood dysregulation and cognitive impairment.

Research has shown that overactivity of 5-HT2a receptors contributes to neuropsychiatric states such as anxiety, depression, and schizophrenia. Conversely, blocking and downregulating hyperactive receptors can alleviate symptoms of these disorders, illustrating their central role in mental health.

I believe the precision afforded by RNA therapeutics will come to represent the most promising advances in psychiatric treatment ever. Traditional medications often, affect multiple receptor systems throughout the brain and body. In contrast, RNA therapeutics targeting the 5-HT2a receptor can achieve a level of specificity previously unimaginable, modulating only the specific neural circuits involved in pathological states while leaving healthy functioning intact.

Modulating the 5-HT2a receptor shows promise in treating psychiatric disorders through a completely different pathway than traditional pharmaceuticals. This receptor plays a crucial role in neuropsychiatric conditions, making it an attractive target for precise therapeutic interventions. By focusing on targeted gene silencing, this approach aims to develop long-lasting treatments that eliminate side effects, providing a safer and more focused alternative to broad-spectrum therapies.

The suffering caused by anxiety disorders and cognitive impairment is immense and often invisible. Patients describe feeling trapped in their own minds, unable to experience the full richness of life or contribute their talents to society. Current treatments, while helpful for many, leave a significant proportion of patients with inadequate relief or troubling side effects. By targeting the 5-HT2a receptor with unprecedented precision, RNA therapeutics offer hope for these individuals—a chance to reclaim their lives and well-being.

Benefits of Targeting 5-HT2a with shRNA

Manipulating the 5-HT2a receptor in animal models has demonstrated remarkable effectiveness in addressing the root causes of disorders such as memory decline and anxiety. This targeted approach not only reduces the symptoms of different illnesses but also significantly decreases side effects compared to conventional treatments.

The use of short hairpin RNA (shRNA) to modulate 5-HT2a receptor expression represents a particularly elegant solution. By precisely controlling the level of receptor expression in specific brain regions, this therapeutic technique offers a promising avenue for enhancing both the safety and efficacy of treatments for neuropsychiatric conditions.

The science behind this approach is compelling, but we must never lose sight of its ultimate purpose: improving human lives. When I consider the potential of RNA therapeutics targeting the 5-HT2a receptor, I envision patients gaining—not just symptom relief—full quality of life—the ability to feel the complete spectrum of human emotions, to think clearly and creatively, and to engage fully with the world around them.

As research in this field progresses, we expect that RNA therapeutics targeting the 5-HT2a receptor will lead to improved patient outcomes and enhanced treatment compliance. By addressing the fundamental molecular mechanisms of disease rather than merely masking symptoms, these next-generation treatments represent a true paradigm shift in our approach to some of the most challenging conditions in psychiatry and neurology.

In the next chapter, we will explore how RNA therapeutics can be precisely delivered to affect these receptor systems in specific brain regions, further enhancing their therapeutic potential while minimizing unwanted effects.

vii. RNA Plasmid Production Procedures

Step-by-step Process of Generating RNA in a Plasmid

The production of RNA via plasmid-based systems represents one of the most powerful and versatile approaches in modern molecular biology. This process, which has evolved significantly since the advent of recombinant DNA technology in the 1970s, enables the precise generation of RNA molecules for both research and therapeutic applications. This chapter provides a comprehensive examination of the techniques, challenges, and quality considerations in RNA plasmid production for RNA neuromodulator development.

Circular plasmid, helical details of active transcription machinery. Rendering.

Historical Evolution of Plasmid Technology

Plasmid-based RNA production has undergone remarkable evolution since the first recombinant plasmids were constructed in the early 1970s. Initially, these circular DNA molecules—naturally occurring in bacteria—were viewed primarily as convenient vectors for cloning and amplifying genes of interest. Early plasmids contained minimal features: an origin of replication, an antibiotic resistance marker, and a simple cloning site.

The development of specialized RNA expression plasmids in the 1980s and 1990s represented a significant advance, incorporating RNA polymerase promoters (such as T7, T3, and SP6) that enabled efficient in vitro transcription. These systems initially focused on producing research-grade RNA for studies of RNA structure and function.

The advent of RNA interference (RNAi) in the late 1990s drove further refinement of plasmid systems, with the development of vectors specifically designed to express short hairpin RNAs (shRNAs) in mammalian cells. These plasmids incorporated polymerase III promoters (such as U6 and H1) optimized for producing small, structured RNA molecules.

Most recently, the emergence of CRISPR/Cas systems and therapeutic RNA applications has spurred the development of highly sophisticated plasmid systems with enhanced features: inducible promoters for temporal control of expression, tissue-specific promoters for spatial control, and advanced selection markers for more efficient generation of stable cell lines.

This historical progression reflects both our deepening understanding of RNA biology and the expanding applications of RNA in basic research, diagnostics, and therapeutics. Today's plasmid systems for RNA production represent the culmination of decades of refinement, offering unprecedented precision and versatility for generating RNA molecules with specific sequences, structures, and modifications.

Conceptualization and Design

The process begins with careful conceptualization and design, where scientists define their objectives and develop the RNA construct that will address specific research or therapeutic goals.

Target Gene Identification

The identification of the target gene is the foundational step in any RNA plasmid production process. This selection must be guided by both scientific understanding of the biological pathway involved and pragmatic considerations regarding the feasibility of modulation.

For RNA neuromodulators, target selection typically focuses on receptors, ion channels, or signaling molecules that play established roles in neuropsychiatric or neurodegenerative conditions. Ideal targets exhibit several key characteristics:

Pathway Validation: Strong evidence linking the target to disease pathophysiology, from multiple lines of investigation including genetic studies, animal models, and human biomarker or imaging data.

Expression Patterns: Targets should be expressed in relevant cell types and brain regions associated with the condition being addressed. Single-cell transcriptomics and spatial transcriptomics have dramatically enhanced our ability to characterize cell type-specific expression patterns in the brain.

Functional Redundancy: Understanding whether compensatory mechanisms might circumvent the therapeutic effect if a single target is modulated. In some cases, targeting multiple related genes may be necessary for efficacy.

Druggability: Some genes are inherently more amenable to RNA-based modulation than others, depending on factors like transcript stability, secondary structure, and sequence uniqueness within the genome.

Bioinformatics analysis plays a crucial role in this process, employing sophisticated algorithms to predict target accessibility, potential off-target effects, and optimal sequences for modulation. These computational approaches draw on diverse datasets including:

1. Transcriptomic databases that provide information on gene expression levels across different brain regions and cell types
2. Structural prediction tools that model RNA secondary structure and accessibility
3. Homology searches to identify potential off-target binding sites
4. Phylogenetic analyses that can highlight evolutionarily conserved regions that may be functionally important

This analysis helps researchers select not just an appropriate gene but the most effective region within that gene for targeting—a critical consideration for successful RNA-based interventions.

RNA Sequence Design

The process of designing the RNA sequence is highly specific to the intended application, with different design principles applying to different RNA-based approaches:

For shRNA (Short Hairpin RNA): Design focuses on creating a stem-loop structure that efficiently enters the cellular RNAi machinery. Key considerations include:

1. A stem region of 19-29 base pairs containing the sequence complementary to the target mRNA
2. A loop region of 6-9 nucleotides that allows the RNA to fold properly

3. Careful positioning of the guide strand to ensure proper loading into the RNA-induced silencing complex (RISC)
4. Strategic placement of mismatches or modifications that can enhance specificity or reduce immune stimulation

For CRISPR Guide RNAs: Design prioritizes target specificity and Cas nuclease compatibility:

1. A 20-nucleotide sequence complmentary to target DNA site
2. Adjacent to an appropriate protospacer adjacent motif (PAM) required for Cas nuclease function
3. Scoring for predicted on-target efficiency based on factors like GC content and secondary structure
4. Comprehensive off-target analysis to minimize unintended genome editing events

For mRNA Therapeutics: Design considerations extend beyond the coding sequence to encompass elements that affect stability, translation efficiency, and immunogenicity:

1. Codon optimization to enhance translation efficiency in human cells
2. 5' and 3' untranslated regions (UTRs) that influence mRNA stability and translation
3. Potential inclusion of modified nucleotides like pseudouridine to reduce immunogenicity
4. Poly(A) tail length optimization for stability and efficiency

Advanced design algorithms now incorporate machine learning approaches trained on experimental data to predict which RNA sequences will exhibit optimal activity while minimizing off-target effects. These tools continuously improve as new data become available, reflecting the iterative nature of RNA therapeutic development.

Oligonucleotide Synthesis

The synthesis of oligonucleotides—short, chemically synthesized DNA fragments that serve as the starting material for many RNA production processes—has been revolutionized by automated solid-phase synthesis technologies. This process, while typically outsourced to specialized commercial providers, remains a critical determinant of downstream success.

Modern oligonucleotide synthesis utilizes phosphoramidite chemistry in a cyclical process that builds DNA strands one nucleotide at a time. The process includes:

1. *Deprotection:* Removing a protective group from the 5' end of the growing chain
2. *Coupling:* Adding a new nucleotide phosphoramidite to the exposed 5' hydroxyl group
3. *Capping:* Blocking unreacted chains prevent sequence errors
4. *Oxidation:* Converting the phosphite linkage to a more stable phosphate bond

This cycle repeats until the desired sequence is complete, at which point the oligonucleotide is cleaved from the solid support and purified.

Quality control of synthesized oligonucleotides is crucial and typically involves:

1. *Mass Spectrometry:* Verifies molecular weight of the synthesized product, confirming that the sequence length is correct
2. *High-Performance Liquid Chromatography (HPLC):* Assesses purity by separating the full-length product from truncated sequences and synthesis byproducts
3. *Capillary Electrophoresis:* Provides high-resolution separation of oligonucleotides based on size and charge, offering additional confirmation of product quality

For longer constructs, overlapping oligonucleotides may be designed and then assembled using enzymatic methods like polymerase chain reaction (PCR) or Gibson Assembly. These approaches enable the construction of complete gene-length sequences from shorter oligonucleotides, which is particularly valuable for RNA therapeutics requiring longer sequences.

Plasmid Selection and Engineering

The selection of an appropriate plasmid vector is critical for successful RNA production, with the choice dependent on both the type of RNA being produced and its intended application.

Vector Selection Criteria

Key considerations in selecting an RNA expression vector include:

Promoter Strength and Specificity: Different applications require different promoters:

1. RNA polymerase II promoters (like CMV or EF-1α) for mRNA expression in mammalian cells
2. RNA polymerase III promoters (U6 or H1) for shRNA expression

3. Bacteriophage promoters (T7, T3, or SP6) for in vitro transcription systems

Backbone Elements: The plasmid backbone must contain essential elements:

1. Origin of replication compatible with the host organism (bacterial, yeast, or mammalian)
2. Selectable marker (antibiotic resistance gene marker)
3. Multiple cloning site (MCS) with appropriate restriction sites

Size Considerations: Smaller plasmids generally transform more efficiently and yield higher copy numbers in bacteria.

Special Features: Depending on the application, vectors may need additional elements:

1. Inducible promoters for controlled expression
2. Tissue-specific promoters for targeted expression
3. Reporter genes monitor transfection or expression efficiency
4. Recombination sites for gateway cloning or site-specific integration

Modern RNA expression vectors often incorporate sophisticated design elements that enhance their utility for specific applications. For shRNA expression, vectors may include flanking sequences that mimic natural microRNA contexts, improving processing efficiency. For mRNA production, vectors might contain optimized 5' and 3' untranslated regions that enhance stability and translation.

Cloning Strategies

The integration of the RNA-encoding sequence into the selected plasmid can be accomplished through several methodologies, each with distinct advantages:

1. *Restriction Enzyme Cloning:* This traditional approach remains valuable for many applications:
2. *Restriction Enzyme Digestion:* Both the plasmid and insert (typically PCR-amplified with primers containing appropriate restriction sites) are cut with restriction enzymes. This creates compatible "sticky ends" that can anneal together.
3. *Ligation:* DNA ligase catalyzes the formation of phosphodiest er bonds between the insert and plasmid, creating a circular recombinant plasmid.

The success of restriction cloning depends on several factors:

1. *Choice* of restriction enzymes
2. *Optimized molar ratios* between vector and insert

3. *Efficient dephosphorylation* to prevent self-ligation

Gibson Assembly and Similar Methods: These newer approaches allow seamless assembly of DNA fragments without restriction enzyme sites:

1. *Design:* Overlapping sequences (15-25 bp) are designed at the ends of adjacent fragments.
2. *Exonuclease Activity:* An enzyme chews back the 5' ends, creating single-stranded overhangs.
3. *Annealing:* Complementary overhangs anneal, joining the fragments.
4. *Polymerase and Ligase Activity:* Gaps are filled and nicks sealed to create intact DNA.

These methods offer advantages for RNA applications where restriction sites might interfere with RNA structure or function. They also enable more complex assemblies involving multiple fragments in a single reaction.

TOPO Cloning: This method exploits the biological activity of topoisomerase I:

1. *Vector Preparation:* Linear vectors with topoisomerase I covalently attached at each end.
2. *Ligation:* PCR products with compatible ends are efficiently ligated into the vector by the bound enzyme.

This approach is particularly valuable for high-throughput applications due to its speed and efficiency.

Gateway Cloning: Based on bacteriophage lambda recombination:

1. *Entry Clone Creation:* The gene of interest is first cloned into an "entry vector."
2. *Destination Vector:* The gene is then transferred to various "destination vectors" via site-specific recombination.

This system is especially useful when the same RNA sequence needs to be expressed under different promoters or in different contexts.

Transformation and Selection

Once the recombinant plasmid is constructed, it must be introduced into host cells for amplification and verification.

Bacterial Transformation

Bacterial transformation—the process of introducing foreign DNA into bacterial cells—can be accomplished through several methods:

1. *Heat Shock Transformation:* This classical method involves:
2. *Cell Preparation:* Treatment of bacteria (typically E. coli) with calcium chloride to make them "competent"—receptive to DNA uptake
3. *Heat Pulse:* A brief exposure to elevated temperature (42°C for 30-90 seconds) creates pores in the cell membrane
4. *Recovery:* Cells are given time to recover in nutrient-rich media before selection.

While relatively simple, heat shock typically achieves transformation efficiencies of 10^6 to 10^8 transformants per microgram of plasmid DNA, sufficient for most standard cloning applications.

Electroporation: This more efficient method uses an electrical pulse to temporarily disrupt the cell membrane:

1. *Cell Preparation:* Bacteria are washed extensively to remove salts that could cause arcing.
2. *Electric Pulse:* A high-voltage electrical discharge creates transient pores in the membrane.
3. *Recovery:* As with heat shock, cells require recovery time before selection.

Electroporation can achieve transformation efficiencies up to 10^10 transformants per microgram of DNA, making it valuable for difficult-to-transform strains or large plasmids.

Several factors influence transformation efficiency:

1. Plasmid size (smaller plasmids transform more efficiently)
2. DNA purity (contaminants like salts or phenol reduce efficiency)
3. Bacterial strain (some are engineered for high competence)
4. Cell density growth phase during competent cell preparation

After transformation, bacteria are plated on selective media containing antibiotics that match the resistance marker in the plasmid. Only cells that have successfully taken up the plasmid will form colonies, creating a straightforward selection system.

Colony Screening

Not all transformants contain the correct plasmid construct; some may harbor empty vectors or plasmids with incorrect inserts. Several screening methods identify colonies with desired construct:

1. *Colony PCR:* This direct screening method involves:

2. *Template Preparation:* A small amount of bacterial colony is picked and added directly to a PCR reaction.
3. *Amplification:* Primers flanking the insertion site amplify the insert if present.
4. *Size Analysis:* PCR products are analyzed by gel electrophoresis to confirm insert presence and size.

This approach provides rapid screening without the need for plasmid isolation, though it can sometimes yield false negatives due to PCR inhibitors in the bacterial culture.

Restriction Digestion Analysis: This traditional method involves:

1. *Mini-prep:* Small-scale plasmid isolation from individual colonies.
2. *Digestion:* Treatment with restriction enzymes that cut at known positions.
3. *Gel Analysis:* The pattern of DNA fragments reveals whether the correct insert is present.

This method provides more definitive results than colony PCR but requires more time and resources.

Blue-White Screening: For plasmids with an appropriate indicator system:

1. *Indicator System:* The plasmid contains the lacZ α-peptide gene with the multiple cloning site inside.
2. Substrate: Plates contain X-gal, which turns blue when cleaved by functional β-galactosidase.
3. *Color Selection:* White colonies indicate insertional inactivation of the lacZ gene, suggesting successful cloning.

While convenient, this method only indicates the presence of an insert, not whether it is the correct one. Modern high-throughput approaches may combine these methods or use automated systems that pick colonies directly into PCR screens or liquid cultures for plasmid isolation.

Plasmid Amplification and Purification

Once colonies with the correct plasmid are identified, larger-scale production and purification are necessary to obtain sufficient material for downstream applications.

Bacterial Culture

The amplification of plasmid DNA relies on bacterial replication machinery to produce multiple copies of the plasmid within each cell, followed by exponential growth of the bacterial population. This process typically follows these steps:

1. *Starter Culture:* A single colony is inoculated into a small volume (2-5 mL) of liquid media containing the appropriate antibiotic. This culture grows for 8-16 hours, reaching stationary phase.
2. *Scale-Up:* The starter culture is diluted (typically 1:100 to 1:1000) into a larger volume of fresh media with antibiotic. For standard laboratory preparations, this might be 50-250 mL, while production-scale preparations might use 1-10 liters or more.
3. *Growth:* The culture is incubated with appropriate aeration (shaking for flasks, controlled aeration for bioreactors) until it reaches the optimal density for harvesting. For high-copy plasmids, this is typically late log phase.

Several factors influence plasmid yield and quality:

1. *Bacterial Strain:* E. coli strains offer various advantages:
   - *DH5α:* High plasmid stability and yield
   - *TOP10:* Improved mRNA stability for certain constructs
   - *Stbl3:* Reduced recombination for unstable constructs
2. *EndA-strains:* Improving DNA quality
3. *Growth Conditions:* Temperature, media composition, and aeration all affect plasmid copy number and bacterial growth. Lower temperatures (30-32°C instead of 37°C) can improve yield for some unstable plasmids.

Plasmid Extraction

The isolation of plasmid DNA from bacterial cultures has evolved from labor-intensive cesium chloride gradient centrifugation to streamlined kit-based methods. Most modern protocols follow a modified alkaline lysis procedure:

1. *Cell Harvesting:* Bacterial cells are pelleted by centrifugation and resuspended in a buffer containing RNase A.
2. Alkaline Lysis: Cells are lysed using a solution containing sodium hydroxide and sodium dodecyl sulfate (SDS), which denatures proteins and genomic DNA.
3. *Neutralization:* Addition of acidic potassium acetate renatures plasmid DNA while causing genomic DNA and proteins to precipitate.

4. *Clarification:* The precipitate is removed by centrifugation or filtration, leaving plasmid DNA in solution.
5. *Purification:* Several methods available for further purification:

- *Silica-Based Columns:* Plasmid DNA binds to silica membranes under high-salt conditions, is washed to remove contaminants, and then eluted in low-salt buffer.
- *Anion Exchange Chromatography*: Plasmid DNA binds to positively charged resins, with contaminants removed through washing steps before elution.
- *Size Exclusion Chromatography:* Separates plasmid DNA from smaller contaminants based on molecular size.

For therapeutic applications, additional purification steps may be necessary:

- *Endotoxin Removal:* Special buffers or resins selectively bind bacterial endotoxins (lipopolysaccharides) that can trigger immune responses.
- *Sterile Filtration:* Removal of any remaining bacteria or particulates through 0.22 μm filters.
- *Gradient Ultracentrifugation:* Separation of supercoiled plasmid from other topological forms for applications requiring highly homogeneous material.

The choice of purification method depends on the intended application, with different purity requirements for research use, in vitro transcription, or clinical applications.

Quality Control

Rigorous quality control is essential to ensure that the purified plasmid meets specifications for identity, purity, and integrity:

- *Restriction Enzyme Digestion:* Confirms the plasmid's identity by producing a specific pattern of fragments.
- *Sequencing* Verifies the exact sequence of the insert and critical regions like promoters and junctions.
- *Endotoxin Testing:* For plasmids intended for transfection or therapeutic use, endotoxin levels must typically be below 0.1 EU/μg DNA.
- *Supercoiled Content:* Assesses the percentage of plasmid in the supercoiled form, which is generally more effective for transfection and in vitro transcription.

Modern analytical techniques like capillary electrophoresis and high-performance liquid chromatography provide more detailed characterization of plasmid preparations, enabling better quality control and batch-to-batch consistency.

In Vitro Transcription (IVT)

In vitro transcription converts the DNA template into RNA, a critical step for many RNA-based applications including the production of mRNA therapeutics and guide RNAs.

Template Preparation

The preparation of a high-quality DNA template is crucial for successful in vitro transcription:

Linearization: For run-off transcription, the plasmid must be linearized downstream of the insert:

1. *Enzyme Selection:* The restriction enzyme must create a clean cut without 3' overhangs, which could lead to undesired transcription products.
2. *Complete Digestion:* Partial digestion results in heterogeneous transcripts, reducing yield and quality.
3. *Purification:* The linearized template must be purified to remove enzymes, buffers, and small DNA fragments that could interfere with transcription.

Alternatively, PCR can be used to generate linear templates with precise termini, which is particularly valuable when specific sequences are required at the 3' end of the RNA.

Transcription Reaction

The core transcription reaction includes several critical components:

RNA Polymerase: Bacteriophage polymerases (T7, T3, or SP6) are most commonly used due to their high processivity and promoter specificity. The choice depends on the promoter present in the template.

Nucleotide Triphosphates (NTPs): High-purity NTPs are essential, with concentrations typically ranging from 1-7.5 mM each. For some applications, modified nucleotides may be incorporated:

1. Pseudouridine or N1-methylpseudouridine to reduce immunogenicity
2. 5-methylcytidine to enhance translation efficiency

3. Cap analogs for co-transcriptional capping of mRNA

Reaction Buffer: Typically contains:

1. Magnesium ions (6-12 mM), which are essential cofactors for polymerase activity
2. Spermidine, helps overcome charge repulsion in the template
3. DTT to maintain reducing conditions
4. Buffer components to maintain optimal pH

RNase Inhibitor: Protects newly synthesized RNA from degradation by contaminating ribonucleases.

DNA Template: Highly purified linearized plasmid or PCR product.

Reaction optimization may involve adjusting component concentrations, incubation times, and temperatures to maximize yield while maintaining product integrity. Modern high-yield systems can produce up to 5 mg of RNA per mL of reaction mixture.

RNA Purification

The purification of transcribed RNA removes template DNA, unused NTPs, and enzymes, yielding a pure RNA product:

Purification Methods: Several approaches are available:

1. *Phenol-Chloroform Extraction:* Traditional method that separates RNA (aqueous phase) from proteins (organic phase), followed by ethanol precipitation.
2. *Silica Column Purification:* RNA binds to silica membranes under high-salt conditions, allowing contaminants to be washed away before elution.
3. *Size Exclusion Chromatography:* Separates RNA from smaller molecules based on size, particularly useful for removing unused NTPs.
4. *HPLC Purification:* Provides higher resolution separation, valuable for critical applications requiring extremely pure RNA.
5. *Gel Purification:* Separates RNA based on size using denaturing polyacrylamide gel electrophoresis, allowing isolation of full-length transcripts. This method, while labor-intensive, provides the highest resolution for removing truncated products.

For therapeutic applications, additional purification steps may be necessary to remove double-stranded RNA impurities, which can trigger immune responses, and to ensure the removal of any remaining endotoxins or host cell proteins.

Quality Control of RNA

Comprehensive quality control ensures that the RNA meets specifications for identity, purity, integrity, and activity:

RNA Quantification

Downsteam applications require accurate measurement of RNA concentration:

1. *Spectrophotometry (UV Absorbance):* Traditional method based on RNA's absorption maximum at 260 nm. While convenient, this method cannot distinguish between full-length RNA and degradation products.
2. *Fluorometry:* Uses RNA-specific dyes like RiboGreen that increase fluorescence upon binding to RNA. This method is less affected by contaminants than spectrophotometry and can be more sensitive for dilute samples.
3. *Microfluidic Analysis:* Platforms like the Agilent Bioanalyzer or Advanced Analytical Fragment Analyzer combine separation and quantification, providing information on concentration and integrity.

Integrity Assessment

The integrity of RNA—the degree to which it remains full-length and undegraded—is critical for functionality:

1. *Denaturing Gel Electrophoresis:* Traditional method that separates RNA by size under denaturing conditions (typically using formaldehyde or urea). A sharp, distinct band indicates intact RNA, while smearing suggests degradation.
2. *Capillary Electrophoresis:* Automated systems like Bioanalyzer provide high-resolution separation, digital data on RNA integrity. The RNA Integrity Number provides a quantitative measure of RNA quality on a scale of 1 (completely degraded) to 10 (perfectly intact).
3. *HPLC Analysis:* Ion-pair reversed-phase HPLC can separate RNA based on length and sequence, providing detailed information on product homogeneity.

Sequence Verification

Confirming the correct sequence of the RNA ensures that it will function as intended:

1. *Reverse Transcription PCR (RT-PCR):* Converts a portion of the

RNA back to DNA for sequencing or other analyses. While not typically used to sequence the entire RNA, it can confirm key regions or junctions.

2. *Mass Spectrometry:* For shorter RNAs (up to ~30 nucleotides), techniques like MALDI-TOF or electrospray ionization mass spectrometry can provide exact mass measurements that confirm sequence composition.
3. *Next-Generation Sequencing:* For comprehensive sequence verification, RNA can be converted to a cDNA library and sequenced. This approach is particularly valuable for complex mixtures or when checking for low-level sequence variants.

Functional Testing

For many applications, functional assays provide the most relevant quality assessment:

1. *Translation Assays:* For mRNA, in vitro translation systems or cell-based reporter assays confirm that the RNA can be translated into protein.
2. *Gene Silencing Assays:* For shRNAs, cell-based assays measuring knockdown of the target gene verify functionality.
3. *CRISPR Activity Assays:* For guide RNAs, in vitro cleavage assays or cell-based editing efficiency tests confirm activity.

These functional tests complement analytical methods by directly assessing whether the RNA will perform its intended biological role.

Formulation and Storage

Proper formulation and storage are essential for maintaining RNA stability and activity:

RNA Stabilization

RNA is inherently unstable due to ubiquitous ribonucleases and its chemical susceptibility to hydrolysis. Several approaches enhance stability:

Buffer Composition: Typically includes:

1. Chelating agents like EDTA to sequester divalent metals that can catalyze RNA degradation
2. pH control (usually slightly acidic, pH 6-6.5) to minimize spontaneous hydrolysis
3. RNase inhibitors for short-term storage

Chemical Modifications: Strategic modifications to the RNA can enhance stability:

1. 2'-O-methyl or 2'-fluoro substitutions at vulnerable positions
2. Phosphorothioate backbone modifications to resist nuclease degradation
3. Terminal modifications to protect 5' and 3' ends

Formulation Additives: Various excipients can enhance stability:

1. Carrier RNA to saturate any contaminating RNases
2. Antioxidants to prevent oxidative damage
3. Cryoprotectants i.e. trehalose, sucrose, freeze-thaw stability

Storage Conditions

The optimal storage conditions depend on the specific RNA and its intended use:

1. *Short-term Storage:* For immediate use (days to weeks), RNA is typically stored at -20°C in RNase-free water or buffer.

2. *Medium-term Storage:* For months of storage, RNA should be kept at -80°C, preferably as an ethanol precipitate or in the presence of stabilizing agents.

3. *Long-term Storage:* For archival storage (years), lyophilization (freeze-drying) provides the greatest stability, especially when combined with appropriate excipients.

4. *Aliquoting:* Dividing RNA into single-use aliquots prevents repeated freeze-thaw cycles, which can accelerate degradation.

For therapeutic applications, stability studies under various conditions are essential to establish shelf life and handling requirements. These typically include real-time stability studies at the intended storage temperature and accelerated conditions to predict long-term stability.

Integration with Delivery Systems

For many applications, particularly therapeutics, the RNA must ultimately be formulated with an appropriate delivery system:

1. *Lipid Nanoparticles (LNPs):* Most advanced RNA therapeutics use LNPs, which encapsulate the RNA in a lipid bilayer with ionizable components that facilitate cellular uptake.

2. *Polymer Nanoparticles:* Various polymers can form complexes with RNA, protecting it from degradation and enhancing cellular delivery.

3. *Viral Vectors:* For some applications, the RNA-encoding plas-

mid may be packaged into viral vectors like adeno-associated viruses (AAVs) for in vivo gene delivery.

The integration of RNA production with these delivery systems requires careful optimization to ensure compatibility and maintain RNA integrity throughout the formulation process.

Future Directions in RNA Plasmid Production

RNA plasmid production continues to evolve, with several emerging trends promising to enhance efficiency, scale, and quality:

1. *Cell-Free Production Systems:* Moving beyond bacterial fermentation to cell-free protein synthesis systems that can produce plasmid DNA without cellular growth, potentially reducing contamination risks and simplifying purification.
2. *Enzymatic DNA Synthesis:* Novel approaches using polymerase-based methods for de novo DNA synthesis could eventually replace chemical oligonucleotide synthesis, enabling longer, more accurate constructs with fewer errors.
3. *Continuous Manufacturing:* Transitioning from batch processes to continuous flow systems that integrate plasmid production, purification, and RNA transcription in a closed system, improving consistency and reducing contamination risks.
4. *Microfluidic Platforms:* Miniaturized reaction systems that enable rapid optimization of conditions and potentially point-of-care production of personalized RNA therapeutics.
5. *AI-Driven Design:* Machine learning algorithms that predict optimal RNA sequences, modifications, and formulations based on the therapeutic target and delivery requirements.

These advances promise to further refine the production of RNA via plasmid systems, enabling more efficient development of RNA neuromodulators and other RNA-based therapeutics with enhanced efficacy, specificity, and safety profiles.

viii. Value Proposition of RNA Therapeutics for Mental Health

From the business development perspective, the emergence of RNA-based therapeutics marks a paradigm shift in mental health treatment, offering a novel approach grounded in rigorous scientific research. While still in its early stages, this field is brimming with potential to transform the lives of the hundreds of millions affected by neuropsychiatric disorders.

Pyramid shaped neurons with radiating basal dendrites. Rendering.

As biotechnology firms navigate the complex regulatory landscape of preclinical and clinical trials, the scientific community is poised for a breakthrough in personalized psychiatric treatment. This new era in mental health care may see the integration of genetic medicine as a frontline therapy—one that addresses the molecular drivers causing dysfunction. Stratagists at a neurotherapeutic startup, conducted an analysis of RNA therapy's potential impact, focusing on two key stakeholder groups to develop hypothetical scenerios: patients seeking effective treatment and healthcare providers responsible for administering care.

A Transformative Approach to Mental Health Treatment

A human-centered investigative approach is essential to evaluating the potential advantages of RNA-based treatments. By combining scientific methodology with an empathetic understanding of patient and provider needs, researchers identified key expectations, challenges, and opportunities. This analysis suggests that RNA therapeutics may not simply offer an incremental improvement but instead represent a fundamental reimagining of mental health care.

The following are the output of a design thinking process, consisting of a series of thought experiments. The following is the work product:

1. RNA Therapeutics Have the Potential to Deliver

1. Enhanced patient outcomes through targeted neuronal receptor modulation for biological intervention.
2. Greater patient satisfaction by reducing side effects and treatment burdens
3. Increased efficiency, profitability for healthcare providers
4. New standard of care fostering trust, long-term engagement

By integrating cutting-edge molecular biology with a deep understanding of neuropsychology, these therapies signify a shift toward precision medicine—one where treatment is tailored to individual neural profiles rather than relying on broad-spectrum pharmacology.

2. Patient Experience: a New Standard of Care

For patients, RNA therapeutics present a revolutionary alternative to conventional treatments like SSRIs and benzodiazepines. Anticipated patient experience:

1. Gradual and sustained therapeutic benefits, eliminating the rollercoaster effect of daily medication
2. Minimal to no adverse side effects, reducing the risk of weight gain, sedation, or emotional blunting
3. A simplified treatment regimen, potentially requiring only a single or infrequent dose rather than lifelong adherence to daily medication
4. Long-term efficacy, allowing patients to experience relief without the burden of constant pharmaceutical intervention

This shift in treatment dynamics has the potential to restore autonomy to patients, improving quality of life while reducing reliance on symptom management strategies.

Healthcare Provider Perspective: Innovation with Practical Benefits

For clinicians, the arrival of RNA therapeutics for mental health may offer an opportunity to redefine treatment protocols with particularl interest in:

1. First-line efficacy, with the potential to offer a more direct and lasting impact on neural function
2. Potential for durable patient improvement, reducing the need for ong-term pharmacological dependency

Certain Risks Remain

1. Unforeseen side effects or complications, necessitating thorough clinical validation
2. Potential revenue loss from fewer patient visits and reduced laboratory monitoring requirements

To address concerns, RNA-based treatments need to demonstrate:

1. Ease of administration, ensuring feasibility in clinical practice
2. Broad insurance coverage, making the therapy accessible to diverse populations
3. A high safety profile, supported by rigorous clinical trials
4. Proven efficacy, validated through long term patient outcomes
5. Comprehensive educational resources, equipping providers with the knowledge needed for informed decision-making

By meeting these criteria, RNA therapeutics can establish themselves as a valuable addition to psychiatric treatment, offering both improved patient care and sustainable integration into healthcare systems.

The Compelling Value Proposition

RNA therapeutics are expected to address the core needs of both patients and healthcare providers by offering:

1. Simplified administration with a potential single-dose or infrequent treatment model
2. Reduced side effects, minimizing patient concerns about adverse reactions

3. Improved patient outcomes, enhancing long-term mental health stability
4. Extended intervals between treatments, reducing the burden of frequent medication adjustments

These factors are needed to enhance confidence in patients and providers, leading to higher satisfaction rates and increased adoption of the therapy. With better treatment responses, positive patient experiences, and consistent referrals, RNA therapeutics present a compelling case for revolutionizing mental health care.

A New Future for Mental Health Treatment

The introduction of RNA-based therapies stands to redefine the treatment landscape, evolving away from symptomatic management toward neurobiological restoration. These state-of-the-art interventions may offer patients lasting relief without the side effects or dependency associated with conventional treatments.

For healthcare providers, they represent an opportunity to offer more effective, scientifically advanced, and ethically responsible care. From a business perspective, the value proposition statements, based on research synthesis, illustrate, post-FDA approval, how the transformative potential of RNA may be received:

"Changing the way individuals seeking treatment are freed from anxiety—without harmful side effects."

Physicians utilizing RNA-based therapeutics may soon be able to deliver a single noninvasive treatment lasting months, freeing patients from anxiety for extended periods without the common drawbacks of traditional medications.

"State of the art. Category-defining, rigorous science and ethics, that's what you get with RNA therapies for mental health."

Patients and clinicians can trust the scientific rigor behind RNA-based treatments, developed by experts at the forefront of neurotherapeutics and genetic medicine.

"The way to be free of anxiety without taking SSRIs or other medications—returning the brain's neural networks to a healthy state."

Unlike SSRIs, RNA therapies may allow patients to regain emotional and cognitive stability without daily reliance on medication.

"Finally, the solution to anxiety and memory loss that we've been waiting for."

By offering noninvasive, long-lasting, and precisely targeted intervention, these therapies have the potential to redefine psychiatric treatment.

A New Era of Mental Health Care

RNA therapeutics may soon offer an innovative, scientifically advanced alternative to traditional psychiatric medications. Addressing mental health disorders through direct modulation, these therapies provide a compelling value proposition for both patients and healthcare providers. With their potential for long-term efficacy, minimal side effects, and simplified administration, Intranasal RNA-based treatments could revolutionize mental health care—ushering in a future defined by precision, personalization, and unprecedented therapeutic impact. Preclinical finings suggest not merely an incremental improvement in treatment options, but a reimagining of mental health care approaches.

As the field progresses, it becomes evident that the development of these therapies transcends conventional medical advancement. It represents the creation of a new therapeutic paradigm that integrates cutting-edge molecular biology with a nuanced current understanding of human neuropsychology. These factors are anticipated to enhance trust and confidence among patients and healthcare providers, potentially leading to high satisfaction rates and consistent referrals.

All Constituents Benefit

RNA therapeutics may provide substantial advantages, including improved patient outcomes, increased customer satisfaction, and enhanced profitability for health care providers. The state-of-the-art treatments will likely foster better patient responses but also drive new referrals and ensure profitable transactions.

Experience and Drug Features

We expect patients will experience a gradual onset of therapeutic effects with no adverse side effects. The ease of a monthly nasal spray may help in achieving high outcomes from the treatment experience and engender trust and satisfaction. The treatments will represent a periodic, high-efficacy intervention without the need for daily medication.

Glossary

Receptor Systems & Neuromodulators

5-HT2a Receptors

A subtype of serotonin receptors that regulate mood, anxiety, and cognition. These receptors are key targets for RNA therapeutics due to their role in modulating serotonin, an important neuromodulator. Selective targeting of these receptors may alleviate symptoms associated with depression and anxiety disorders without the systemic effects of traditional medications.

Adrenergic Receptors

Receptors that respond to the neuromodulators adrenaline and noradrenaline (epinephrine and norepinephrine). These play crucial roles in the stress response and arousal. RNA therapeutic approaches targeting these receptors may help reduce stress-related symptoms and regulate autonomic nervous system function with greater precision than conventional treatments.

AMPA Receptors

A subtype of glutamate receptors critical for fast synaptic transmission and neuroplasticity. RNA-based modulation of AMPA receptors shows promise for enhancing cognitive function and treating neuropsychiatric disorders through targeted regulation of glutamate, a primary excitatory neuromodulator in the brain.

Cannabinoid Receptors

Receptors (primarily CB1 and CB2) that respond to endocannabinoids, an important class of neuromodulators that influence pain perception, appetite, and emotional regulation. RNA therapeutics targeting these receptors may offer novel approaches to pain management and mood disorders without the psychoactive effects associated with traditional cannabinoids.

Dopamine Receptors

A family of receptors (D1-D5) that respond to dopamine, a key neuromodulator involved in reward processing, motivation, and motor control. Dysregulation is associated with conditions including schizophrenia, Parkinson's disease, and addiction. RNA therapeutics can potentially offer more selective modulation of specific receptor subtypes than conventional pharmaceuticals.

GABA Receptors

The primary inhibitory receptors in the central nervous system that respond to gamma-aminobutyric acid (GABA), a crucial inhibitory neuromodulator. They reduce neuronal excitability and prevent excessive activity. RNA therapeutics targeting GABA receptor expression offer potential for anxiety treatment with fewer side effects than traditional GABAergic drugs like benzodiazepines.

Glutamate Receptors

The primary excitatory receptors in the brain that respond to glutamate, the major excitatory neuromodulator. They play essential roles in synaptic plasticity, learning, and memory. Dysregulation is linked to conditions including schizophrenia and neurodegenerative diseases. RNA therapeutics may provide more selective modulation of specific receptor subtypes.

Muscarinic Acetylcholine Receptors

A class of receptors that respond to acetylcholine, a neuromodulator involved in learning, memory, and motor control. RNA-based approaches targeting these receptors may provide novel therapeutic avenues for cognitive enhancement and treatment of neurodegenerative disorders like Alzheimer's disease.

Nicotinic Acetylcholine Receptors

Another class of acetylcholine receptors involved in neurotransmission and cognitive function. RNA therapeutics targeting these receptors could potentially enhance cognitive performance with fewer side effects than current cholinergic treatments.

NMDA Receptors

N-methyl-D-aspartate receptors are glutamate receptors that play critical roles in synaptic plasticity and memory function. RNA therapeutics selectively targeting NMDA receptor expression may offer novel approaches for treating depression and Alzheimer's disease by modulating glutamatergic signaling in specific neural circuits.

Opioid Receptors

Receptors involved in pain regulation and reward perception. RNA therapeutics targeting these receptors could potentially provide pain relief without the addictive properties and respiratory depression associated with traditional opioid medications.

Brain Networks & Structures

Corticofugal Network

A complex network of neural connections originating from the cerebral cortex and extending to various subcortical regions. RNA therapeutics delivered to this network may help regulate information flow between cortical and subcortical structures, potentially addressing conditions involving disrupted connectivity.

Corticofugal Neurons

Long-range projection neurons that connect the cortex with subcortical areas, playing crucial roles in motor control and sensory processing. These neurons represent potential targets for RNA therapeutics designed to modulate specific neural pathways.

Default Mode Network (DMN)

A neural network active during rest and self-reflective thinking. Dysregulation of the DMN is associated with conditions like depression and anxiety. RNA therapeutics could potentially normalize DMN activity through targeted modulation of neuromodulator systems within this network.

Dorsal Attention Network (DAN)

A neural network responsible for goal-directed attention and visual-spatial processing. Dysfunction can lead to attention deficits and cognitive impairments. RNA therapeutics targeting neuromodulator systems within this network show promise for addressing attention-related disorders.

Hippocampal Hyperactivity

Excessive neural activity in the hippocampus, a brain region crucial for memory formation. This hyperactivity is associated with anxiety disorders and cognitive impairment. RNA therapeutics may help normalize hippocampal activity by targeting specific neuromodulator receptors in this region.

Limbic System

A set of brain structures including the hippocampus and amygdala that regulate emotions, memory, and motivation. The rich blood supply and proximity to the nasal cavity make this system an ideal target for RNA therapeutics delivered via nasal administration.

Prefrontal Cortex Network (PFC)

A neural network involved in executive functions, decision-making, and attention. RNA therapeutics targeting neuromodulator systems within the PFC show potential for treating conditions involving executive dysfunction.

Salience Network (SN)

A neural network responsible for detecting and filtering significant stimuli and coordinating appropriate responses. RNA therapeutics may help regulate this network in conditions like anxiety where salience attribution is disrupted.

Cellular Components

Astrocytes

A type of glial cell in the central nervous system that provides support for neurons and helps maintain the blood-brain barrier. RNA therapeutics may target astrocytes to indirectly modulate neuronal function or to enhance delivery of therapeutic agents to neurons.

Glial Cells

Support cells in the brain, including protoplasmic astrocytes in gray matter and fibrous astrocytes in white matter. They support neuronal function, maintain the blood-brain barrier, and regulate blood flow. RNA therapeutics may leverage these cells for improved drug delivery or to address glial dysfunction in neuropsychiatric disorders.

Neuronal Progenitors

Cells that can differentiate into various types of neurons. RNA therapeutics may stimulate neuronal progenitor activity to promote neurogenesis and neural repair in conditions involving neuronal loss or dysfunction.

Rostral Migratory Cells

Cells that move from the subventricular zone to the olfactory bulb, promoting neurogenesis. RNA therapeutics may enhance this process to support neural regeneration in neuropsychiatric conditions.

Mental Health Conditions

Anxiety Spectrum Disorders

A group of mental health conditions characterized by excessive fear and apprehension, including generalized anxiety disorder (GAD), panic disorder, and social anxiety disorder. RNA therapeutics offer potential for more targeted treatment of these conditions by addressing specific neuromodulator imbalances.

Mild Cognitive Impairment (MCI)

A decline in cognitive ability beyond normal aging but not significantly interfering with daily functioning. RNA therapeutics targeting memory-related neuromodulator systems show promise for slowing or reversing MCI progression.

RNA Therapeutic Approaches

CRISPR/Cas9

A genome editing technique allowing precise DNA modifications using guide RNA and the Cas9 enzyme. This technology can be adapted for RNA-based therapeutic applications in mental health by targeting specific genes involved in neuromodulator regulation.

Gene Silencing

The process of inhibiting or suppressing gene expression. RNA therapeutics can use this approach to downregulate overactive genes contributing to neuropsychiatric symptoms.

Gene Therapy

A technique involving manipulation of genes in a patient's cells by adding, deleting, or altering genetic material. RNA-based gene therapies for mental health conditions target genes involved in neuromodulator production or receptor function.

RNA Interference (RNAi)

A biological process where RNA molecules degrade specific mRNA molecules, disrupting gene expression. This approach is used in RNA therapeutics to suppress targeted genes involved in neuropsychiatric disorders.

RNA Sequence Design

The process of creating therapeutic RNA sequences involving careful identification of specific gene regions to minimize unintended effects while optimizing stability and efficacy. This is a critical step in developing RNA therapeutics for mental health.

Short Hairpin RNA (shRNA)

A specific RNA sequence that forms a tight hairpin structure, used to efficiently silence target genes in RNA interference applications. shRNAs are being developed to target genes involved in neuropsychiatric disorders.

Synthetic Biology

The manipulation of biological processes for novel applications, such as developing RNA-based treatments for complex neuropsychiatric disorders. This field integrates engineering principles with molecular biology to create innovative therapeutic approaches.

Delivery Methods & Clinical Implementation

Blood-Brain Barrier (BBB)

A specialized barrier separating circulating blood from the brain and extracellular fluid in the central nervous system. It protects the brain from harmful substances while also presenting a challenge for delivering therapeutic agents. RNA therapeutics often incorporate specialized delivery mechanisms to cross this barrier effectively.

Nasal Administration System

A non-invasive method for delivering medication through the olfactory system, bypassing the blood-brain barrier and delivering therapeutic molecules directly to the brain. This approach is particularly promising for RNA therapeutics targeting mental health conditions.

Viral Vector

A commonly used tool in gene therapy for transporting genetic material into cells. Viral vectors serve as safe and effective vehicles for delivering therapeutic RNA to target brain regions in the treatment of neuropsychiatric disorders.

Research & Development Tools

Behavioral Assays

Experiments evaluating therapy effects on animal behavior, particularly regarding anxiety and memory. These tests provide crucial insights into the therapeutic efficacy of RNA-based treatments before clinical trials.

Immunohistochemistry (IHC)

A laboratory technique using antibodies to identify and analyze protein distribution in tissue samples, helping assess effects of RNA therapeutics on target protein expression in the brain.

Multi-Electrode Array (MEA)

A research tool for assessing the effects of genetic modifications and drug therapies on neuronal function by monitoring electrical activity. MEAs help evaluate how RNA therapeutics affect neural network activity.

Polymerase Chain Reaction (PCR)

A scientific technique to amplify small DNA or RNA sections, widely used in research, diagnostics, and forensics. PCR is essential for developing and testing RNA therapeutics.

Clinical Development & Commercialization

Contract Manufacturing Organizations (CMOs)

Organizations that manufacture pharmaceutical products according to precise standards, ensuring consistent production and regulatory compliance. CMOs play a crucial role in scaling up production of RNA therapeutics.

Contract Research Organizations (CROs)

Specialized businesses providing research services including preclinical investigations and clinical trials. CROs offer expertise and resources essential for advancing RNA therapeutics from laboratory to clinic.

Food and Drug Administration (FDA)

The regulatory agency that approves the marketing and sale of new treatments based on safety and efficacy evidence. FDA approval is a critical milestone in bringing RNA therapeutics to market.

Good Manufacturing Practice (GMP)

Guidelines and regulations ensuring product quality and safety during manufacturing. GMP compliance is essential for RNA therapeutics production.

Intellectual Property (IP)

Legal rights granted to creators or inventors of novel products or processes, including patents for innovative pharmaceuticals and delivery mechanisms. Strong IP protection is crucial for commercial development of RNA therapeutics.

Investigational New Drug (IND)

A formal FDA request for permission to conduct human clinical trials, including extensive preclinical research data and a detailed clinical trial plan. IND approval is a key regulatory milestone for RNA therapeutics.

Market Overview

The process of launching a new medication involving marketing, sales, distribution, and education of healthcare professionals and patients. Effective market introduction is essential for successful implementation of RNA therapeutics.

Phase I Clinical Trials

Initial studies of new medicinal approaches focusing on safety, recommended dosage, and side effects. These trials represent the first human testing of RNA therapeutics.

Phase II Clinical Trials

Studies evaluating a medicine's efficacy and safety in a larger patient cohort, determining optimal dosage and therapeutic efficacy. These trials provide crucial data on RNA therapeutics' effectiveness for specific conditions.

Phase III Clinical Trials

Extensive studies validating a medicine's efficacy, monitoring side effects, and comparing it with existing therapies. Successful Phase III trials are typically required for FDA approval of RNA therapeutics.

Regulatory Achievements

Compliance with regulatory standards and guidelines, including IND approval, advancement through clinical trial phases, and marketing authorization. These milestones mark progress in bringing RNA therapeutics to market.

Safety and Toxicology of Substances

Research evaluating potential adverse effects of treatments to ensure new medication safety before human trials. These studies provide crucial information on potential adverse reactions and toxicity of RNA therapeutics.

Therapeutic Index

The ratio of a medication's therapeutic dose to its toxic dose, indicating its safety margin. A high therapeutic index is desirable for RNA therapeutics to minimize side effects while maintaining efficacy.

Other Related Terms

Epigenetics

The study of gene expression changes occurring without alterations to the underlying DNA sequence. These changes can be influenced by environmental factors, personal choices, and medical conditions. RNA therapeutics may target epigenetic mechanisms to normalize gene expression in neuropsychiatric disorders.

Ligation

The process of joining two or more molecules by creating a covalent bond. This technique is used in the development of certain RNA therapeutic constructs.

Memory-Enhancing Cognitive Recall

Interventions designed to improve memory function, including medications, gene therapies, and lifestyle changes. RNA therapeutics targeting memory-related neuromodulator systems represent a promising approach in this category.

Index

Symbols

D

E

L

M

Q

R

S

T

W

www.ingramcontent.com/pod-product-compliance
Lightning Source LLC
Chambersburg PA
CBRC090022130526
44590CB00038B/140

* 9 7 8 1 8 9 2 6 5 4 4 2 7 *